AUTHENTIC OPPORTUNITIES FOR WRITING ABOUT MATH
in Middle School

Teach students to write about math so they can improve their conceptual understanding in authentic ways. This resource offers hands-on strategies you can use to help students in grades 6–8 discuss and articulate mathematical ideas, use correct vocabulary, and compose mathematical arguments.

Part One discusses the importance of emphasizing language to make students' thinking visible and to sharpen communication skills, while attending to precision. Part Two provides a plethora of writing prompts and activities: Visual Prompts; Compare and Contrast; The Answer Is; Topical Questions; Writing About; Journal Prompts; Poetry; Cubing and Think Dots; RAFT; Question Quilts; and Always, Sometimes, Never. Each activity is accompanied by a clear overview plus a variety of examples. Part Three offers a crosswalk of writing strategies and math topics to help you plan, as well as a sample anchor task and lesson plan to demonstrate how the strategies can be integrated.

Throughout each section, you'll also find Blackline Masters that can be downloaded for classroom use. With this book's engaging, standards-based activities, you'll have your middle school students communicating like fluent mathematicians in no time!

Tammy L. Jones has taught students from first grade through college. Currently, she is consulting with individual school districts in training mathematics teachers on effective techniques for being successful in the mathematics classroom, supporting mathematics instruction, and Science, Technology, Engineering,

Mathematics (STEM) integrations. She is co-author of two book series published with Routledge: *Strategies for Common Core Mathematics* and *Strategic Journeys for Building Logical Reasoning.*

Leslie A. Texas has over 20 years of experience working with K–12 teachers and schools across the country to enhance rigorous and relevant instruction. She believes that improving student outcomes depends on comprehensive approaches to teaching and learning. She is co-author of two book series published with Routledge: *Strategies for Common Core Mathematics* and *Strategic Journeys for Building Logical Reasoning.*

Also Available from Tammy L. Jones and Leslie A. Texas
(www.routledge.com/k-12)

Strategic Journeys for Building Logical Reasoning, K–5:
Activities Across the Content Areas

Strategic Journeys for Building Logical Reasoning, 6–8:
Activities Across the Content Areas

Strategic Journeys for Building Logical Reasoning, 9–12:
Activities Across the Content Areas

Strategies for Common Core Mathematics:
Implementing the Standards for Mathematical Practice, K–5

Strategies for Common Core Mathematics:
Implementing the Standards for Mathematical Practice, 6–8

Strategies for Common Core Mathematics:
Implementing the Standards for Mathematical Practice, 9–12

AUTHENTIC OPPORTUNITIES FOR WRITING ABOUT MATH
in Middle School

Prompts and Examples for Building Understanding

Tammy L. Jones and Leslie A. Texas

Routledge
Taylor & Francis Group

NEW YORK AND LONDON

Cover images: @ Getty Images

First published 2025
by Routledge
605 Third Avenue, New York, NY 10158

and by Routledge
4 Park Square, Milton Park, Abingdon, Oxon, OX14 4RN

Routledge is an imprint of the Taylor & Francis Group, an informa business

ISBN: 978-1-032-44931-9 (hbk)
ISBN: 978-1-032-44785-8 (pbk)
ISBN: 978-1-003-37458-9 (ebk)

DOI: 10.4324/9781003374589

Typeset in Warnock Pro
by codeMantra

Access the Support Material: https://resourcecentre.routledge.com/books/9781032447858 or visit https://resourcecentre.routledge.com and search for the book's ISBN, title or authors.

Special thanks to Pixaby, Padowan Graph, Sketchpad, WordArt, and Geometer's Sketchpad for certain images used in this book.

Online Resources

Several of the resources in this book are available online as free downloads so you can print them for classroom use. To access them, find the book at the url below and search for this book's ISBN, title, or authors. Note that you will be asked to provide information from the book before you can obtain the downloads.

https://resourcecentre.routledge.com/

You can also follow this direct link: https://resourcecentre.routledge.com/books/9781032447858

Contents

Contents

Meet the Authors

Collectively Tammy and Leslie have almost 45 years of classroom experience teaching in elementary, middle, high school, and college. This has included urban, suburban, rural, and private school settings. Being active members of their professional organizations has allowed them to continually grow professionally and model lifelong learning for both their students and their peers. In their 30-plus years of combined consulting work, they have had opportunities to work with teachers and students from kindergarten through college level. This work has spanned almost all 50 states. Their work has included helping to develop standards and curriculum at the state level as well as implementing curriculum and best practice strategies at the classroom level. One of the things that set Tammy and Leslie apart as consultants is their work with classroom teachers, modeling and offering continued support throughout the year to build capacity at the building and district levels. Tammy and Leslie co-authored the 2013 series from Eye On Education/Routledge-Taylor & Francis Group, Strategies for Common Core Standards for Mathematics: Implementing the Standards for Mathematical Practice (Grades K–5, 6–8, and 9–12), and the 2017 series from Routledge-Taylor & Francis Group, Strategic Journeys for Building Logical Reasoning: Activities Across the Content Areas (Grades K–5, 6–8, and 9–12).

An educator since 1979, **Tammy L. Jones** has worked with students from first grade through college. Currently, Tammy is consulting with individual school districts in training teachers on strategies for making content accessible to all learners. Writing integrations and literacy connections are foundational in everything Tammy does. Tammy also works with teachers on effective techniques for being successful in the classroom. As a classroom teacher, Tammy's goal was that all students understand and appreciate the content they were studying; that they could read it, write it, explore it, and communicate it with confidence; and that they would be able to use the content as they need to in their lives. She believes that logical reasoning, followed by a well-reasoned presentation of results, is central to the process of learning and that this learning happens most effectively in a cooperative, student-centered classroom. Tammy believes that learning is experiential and in her current consulting work creates and shares engaging and effective educational experiences.

Leslie A. Texas has over 25 years of experience working with K–12 teachers and schools across the country to enhance rigorous and relevant instruction. She believes that improving student outcomes depends on comprehensive approaches to teaching and learning. She taught middle and high school mathematics and science, and has strong content expertise in both areas. Through her advanced degree studies, she honed her skills in content and program development and student-centered instruction. Using a combination of direct instruction, modeling, and problem-solving activities rooted in practical application, Leslie helps teachers become more effective classroom leaders and peer coaches.

We would like to give a special thanks to Trevor Styer for his work to ensure the graphics used throughout the series were high quality and reproducible for classroom use.

Preface

A Note to Our Readers

Our previous two book series, Strategies for Common Core Mathematics: Implementing the Standards for Mathematical Practice (Grades K–5, 6–8, and 9–12) and Strategic Journeys for Building Logical Reasoning: Activities Across the Content Areas (Grades K–5, 6–8, and 9–12), provided a set of strategies and sample tasks that teachers could implement across the curriculum to engage students at a deeper cognitive level required by the rigorous college and career ready standards.

When we took on writing this new series, we asked ourselves: What is it that teachers want and would support students in becoming better communicators of mathematics? During training with teachers on our other two series, we often were asked how teachers could get more classroom-ready materials, such as questions and writing prompts, that would support their work with students on writing and reading mathematics. Therefore, we wanted to create a collection of items for educators that would be practical and versatile, easy to implement, and yield results.

For the student, we created a collection of visual prompts that provide opportunities to engage in mathematics through looking at pictures of and from the world. There is an assortment of examples supporting the academic vocabulary associated with each math topic. Also included are ready-to-use writing prompts covering a variety of topics across the grade bands. Sets of nontypical questions are provided to promote developing a deeper understanding of mathematics. Examples of various writing styles, including creative writing, meet the needs and interests of a diverse classroom.

For educators, it is important to understand students can only become comfortable (and proficient) communicating about mathematics by practicing it regularly. Today's high-stakes assessments require students to understand mathematics in context and to explain their reasoning behind strategies and solutions. There are enough strategies included to incorporate often (daily/weekly). Using these prompts and tasks is easy once the teacher has determined the instructional goals and targeted standards for implementation. There is teacher autonomy in implementing, but the prompts and tasks are ready to be used immediately.

These are great strategies for providing a variety of ways to engage students in mathematical discourse. The materials are versatile in use as handouts, visual displays, gallery walks, electronic documents, etc. The crosswalk shows examples by mathematical topic as well as by type of writing. Teachers will find strategies for authentically integrating different writing techniques in the mathematics classroom, including creative writing. A sample lesson incorporating a number of these prompts and examples is included along with unique strategies and examples for differentiation in the mathematics classroom.

PART ONE

Why Writing in Math Matters

CHAPTER 1

Purposeful Writing: Intentional Design

Communication is essential in expressing ideas clearly and effectively. Language serves as a framework for that communication. Mathematics is often said to have its own language using symbols in addition to words. Combining mathematical language with written/spoken language can often provide deeper insight into how information is being processed, connections that are being made, conclusions drawn, etc. This data is important in assessing understanding as well as moving thinking further.

This book will look at how writing can be used in

1. Making student thinking visible – formative assessment
2. Building communication skills while attending to precision – construct a viable argument and critique the reasoning of others (Standard of Mathematical Practice 3) and attend to precision (Standard of Mathematical Practice 6)
3. Establishing authentic reasons for writing, not just so we can say we did write in math.

There are seven opportunities for writing described below. These served as a guideline and informed the choices made regarding the types of writing included in this series (Jones & Texas, 2017).

DOI: 10.4324/9781003374589-2

- **Making Meaning** – understanding the question posed and identifying the given and needed information necessary to proceed
- **Showing Evidence** – using facts and/or data to support one's argument/hypothesis/work
- **Reflecting** – being metacognitive with respect to strategies and/or processes
- **Inquiry** – creating questions to drive investigation and/or research
- **Educating** – informing others in various forms/purposes (persuasive, descriptive, expository, and narrative)
- **Creating Ideas** – brainstorming/free writing to begin framing ideas
- **Producing Products** – using products to convey a message depending on audience and purpose (research papers, proposals, brochures, essays, public service announcements, etc.)

Making Student Thinking Visible: Formative Assessment

For teachers to elicit the evidence of student understanding and provide feedback that moves the learning forward, students must be able to make their thinking visible. Many students struggle to organize their thoughts and capture their thinking on paper. Starting with a straightforward tool such as the Think-Write-Pair-Share (Jones & Texas, 2017) allows student specific guidance in where to begin writing. A blank piece of paper can mean "I don't know" or "I don't care." It is an important distinction and providing tools for students to support the "I don't know" is critical in building their capacity to help themselves. In addition, this intentional emphasis on writing highlights the importance of being a good partner by bringing something to the table when coming together to discuss ideas. Below is one example of capturing this thinking.

Think, WRITE, Pair, Share

<u>Think</u> about…

<u>WRITE</u> about what questions come to mind in the area below.

<u>PAIR</u> with your partner and discuss what each of you wrote.

Be prepared to **<u>SHARE</u>** with the whole group.

Using with a Rich Task

The Think-Write-Pair-Share tool can be used to help students organize their thoughts around a rich task. The task can be embedded so students can stay focused while engaging in the process. The "write" component gives very specific guidance to support students whether they understand the problem or not. It also provides stems for students to consider any time they are engaged in solving a problem.

Think-Write-Pair-Share	
Think Think about the problem. INSERT TASK HERE	**Write** Write by doing one of the following: If you can solve, choose a strategy and solve. If you cannot solve… ❏ Write all facts you know about the problem ❏ Write anything you know related to the concept addressed in the problem ❏ Write questions you have about the problem
Workspace	
Pair Pair with a partner and take turns discussing your strategies and solutions. Use this space to record strategies that were different from yours.	**Share** Share various strategies and solutions with the group. Use this space to record strategies that were different from those of you and your partner.

See Section 4 for Sample Lesson Plan using this tool.

Building Communication Skills
While Attending to Precision

The following problem-solving process and graphic organizer can be used to assist students in making sense of problems (Standard of Mathematical Practice #1) as well as decontextualizing and contextualizing word problems (Standard of Mathematical Practice #2). The process also requires students to construct viable arguments (Standard of Mathematical Practice #3) as they formulate their own ideas about the meaning of the problem and make predictions about the outcome. Once a solution is obtained, students compare to the prediction to determine the reasonableness of the solution. By giving students explicit steps to unpack the problem, they begin the process with minimal to no teacher guidance and complete the initial steps. This eliminates the blank piece of paper or the famous "I don't know" answer. Using a consistent process over time with students will assist them in becoming better problem solvers. While this process may not always "fit" every problem, it does help students develop a systematic approach to finding the "entry point" into various tasks. (Texas & Jones, 2013)

The process is like the Three Reads strategy in that it asks students to read the problem more than once. The first time they read it in its entirety to understand the context of the problem. Steps #1 and #2 then ask students to reread specific sentences as they decode the text and make sense of the problem. See below for an explanation and how the graphic organizer is used to capture the process.

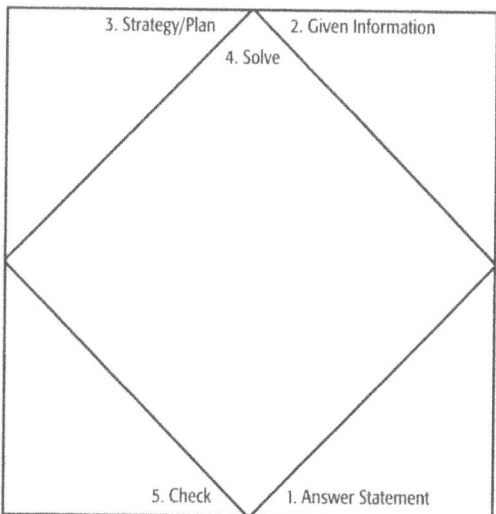

This organizer can be enlarged and copied onto paper for students. It can also be created by folding a piece of paper at each end as if making a paper airplane. Once opened, it will be partitioned as above.

1. **Answer Statement**

   a. The question usually appears as the last sentence of the problem. Students can cover the other information and focus on the last line to determine what the problem is asking. (If the question is not here, students can check each preceding line until it is found.)

   b. Students write the question as an answer statement and leave a blank for the solution. Translating from a question to an answer statement can be challenging for some students. Practicing verbally asking and answering the question can assist in this process.

   c. Remind students to include the appropriate units for the context of the problem.

   d. The answer statement is a critical component and should be practiced even when not using this organizer. It ensures students understand the question being asked as well as guarantees they will answer the question posed if the problem contains multiple steps. Developing this habit promotes its transfer to testing situations and is particularly important when answering constructed response questions.

2. **Given Information**

   a. Students use the same process of viewing each sentence separately, covering everything else.

   b. Students determine and record the relevant information from the problem.

3. **Strategy/Plan**

   a. Students use this space to state additional ideas they have about the problem, such as other information they know about the problem, possible strategies for getting started, estimations for the solution, constraints, or predictions.

   b. This is the section that allows students to formulate their own ideas about the problem and provides a place for them to create their own meaning about what is being asked.

   c. Determining an estimate also provides a context for checking for reasonableness of the solution.

   d. This step also allows students to become strategic problem solvers rather than impulsive ones by requiring them to consider the various strategies available and then determine which might be the most efficient to use in the given situation.

   e. Many students are not versatile in the various problem-solving strategies available. Creating a Strategy Wall can be useful to building the students toolkit. See p. 30 for more information on Strategy Walls.

4. **Solve**

   a. Students select a strategy (translate verbal statements into mathematical statements, draw a picture, make a table, etc.) and solve.
   b. Students can compare their solution to the estimation to determine the reasonableness of their answer.

5. **Check**

   a. Students check their answers by substitution or by using another method to justify.
   b. This is also a good time to strategically partner students who used different strategies. Students can coach each other in the use of their strategy.
   c. Once the answer has been checked, students write the answer in the blank from Step #1.

Emphasis on Process over Solution

The purpose of any problem-solving process is to encourage students to think about the problem before impulsively jumping ahead to solving. It also encourages them to read and understand before assuming what is expected. To reinforce this point, students can be given a set of problems in which they are asked to complete the initial steps but not to solve. This allows the focus to be on making sense of the problem and planning before executing. If on a teaching team, this assignment can be completed in the English Language Arts (ELA) classroom since it involves decoding text and prewriting skills. Once students have completed these initial steps, take away the problem set and have students complete the process by solving, checking, and answering the question. By not having access to the original problems, this will serve as an assessment of the initial steps. If students can complete the work, then the information gathered was sufficient. If not, it reveals key components that were overlooked.

Update and New Information

Since the publication of *Strategies for Common Core Mathematics: Implementing the Standards for Mathematical Practice* (Texas & Jones, 2013), many teachers have asked why the graphic organizer begins at the bottom right rather than the top left. There are two reasons it is organized in this manner. The first was in response to how the brain works when asked to attend. To focus and not just mindlessly record answers in a familiar sequence/order, the brain must consciously engage with the organizer, and therefore, students are more intentional with the process. The second reason was in thinking ahead to when the tool would no longer be needed once the process was internalized. Most mathematics problems begin to be solved at the top left of the problem and then worked down to the bottom right where the solution usually is completed. This organizer begins with the end in mind (bottom right) and then comes full circle with the final answer.

The problem-solving process graphic organizer can be adapted to meet the needs of teachers and students and even eliminated as an organizer for students who internalize the process and no longer need the scaffold. Below is an example of a graphic organizer that was modified from the original. The Problem-Solving Process (Scaffolded) table contains scaffolds where there is a list of possible concepts/strategies for students to select as they build their toolkit. NOTE: The choices given here are general for illustration purposes and would be intentionally crafted for the specific unit in which it was being utilized. The Problem-Solving Process table has the supports removed.

Problem-Solving Process (Scaffolded)

The problem is asking me to... Answer Statement:	I know...
Topic/concept this is related to... Examples: Combining like terms Solving system of equations Unit rates Proportional relationships	**Strategy for solving...** Draw a picture Guess and check Work backwards Use the standard algorithm Etc...
Solve (Show work here)	
This solution means...	

Problem-Solving Process

The problem is asking me to... Answer Statement:	I know...
Topic/concept this is related to...	Strategy for solving...

Solve (Show work here)

This solution means...

Questioning: A Tool for Promoting Communication

There are four opportunities for questioning students while working through the problem-solving process (Jones & Texas, 2017).

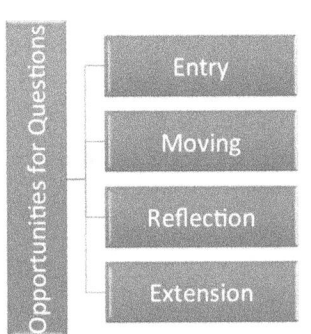

Entry: questions for students having difficulty getting started (Steps #1 and #2)

Moving: questions for places where students could get stuck (Steps #2–#4)

Reflection: questions for students to use for metacognition after completing the problem/issue (Steps #4 and #5)

Extension: questions for students to engage in higher-order thinking skills with respect to the same concept and/or problem (after completing Step #5 and returning to Step #1)

These opportunities allow teachers to develop task-specific questions that can be used to support students as they are working through the process. In using these with students, it was noted that the first two opportunities occurred while students were in the middle of the process and the last two, reflection and extension, could occur once the process has been completed. Therefore, rather than viewing as four opportunities, they were condensed into two: I'm stuck and I'm done.

I'm Done/I'm Stuck

Four Question Types

1. **Entry Questions**
2. **Moving Questions**

3. **Reflection Questions**
4. **Extension Questions**

Task-specific questions can be generated and provided to students as needed, or they can be taught to access them on their own. If stuck, they can retrieve the appropriate questions that will allow them to move forward. For students who finish early, the done questions will be used to have them go deeper with the task rather than be assigned additional work, which is oftentimes seen as busy work. **See Section 4 in Chapter 4 for examples.**

Establishing Authentic Reasons for Writing

Incorporating literacy across the curriculum has long been an emphasis in mathematical classrooms. Initially, this involved having students put into words how they solved the problem alongside the mathematical steps. Fortunately, the redundancy of this request was soon realized. The mathematics itself clearly articulated what students did to solve the problem. Therefore, students were asked to write about their thinking rather than what they did. For example, if solving an equation, students were asked to write about the properties of equality used to explain why they did rather than what they did.

Part II introduces ten different strategies that can be used to provide authentic opportunities for students to write about math. Explanations of the strategies as well as content-specific examples have been provided to make these ready-to-use in the classroom. In addition, each provides the opportunity for differentiation. Below is a brief description of each:

Visual Prompts: pictures and images to initiate thoughts and discussions

Compare/Contrast: academic vocabulary word pairs to deepen understanding

The Answer Is...: giving an answer for which there could be multiple questions posed

Topical Questions: set of questions whose stems promote mathematical discourse

Writing About: using word clouds with academic vocabulary to write about a specific topic

Journal Prompts: assortment of ideas to engage students in journaling about mathematics

Poems: collection of ideas to engage linguistic learners in expressing mathematical thought

Cubing/Think Dots: activities for independent learning

Question Quilts: an alternative way to present questions and provide student agency

RAFT: creative writing opportunity

Always, Sometimes, and Never: alternative way to view statements that promote critical thinking

Writing Prompts

CHAPTER 2

Visual Prompts

The visual prompts given here are actual photographs taken by the authors. These are different from what is known as "visual mathematics" which usually references the various visual representations in mathematics. The picture prompts harken back to *Mister Rogers' Neighborhood* picture segments. These pictures help our students see the mathematics that is all around them. They also offer opportunities for students to engage in authentic mathematical communication.

The following collection of photographs can be used as journal prompts, discussion starters, or bell ringers, or for centers, small groups, or learning stations.* These pictures provide opportunities for students to engage in mathematics through looking at pictures of and from the world. As a starting point, have students free write what they see and describe it. This could be facilitated much like the *Notice and Wonder* prompts that the National Council of Teachers of Mathematics has brought to the forefront in the past few years.

Middle school students can name the geometric shapes they see used, the types of numbers they see, the type of function that might mirror parts of the pictures, and the mathematical topic that the image might conjure up, such as the bear outside the Denver Convention Center and scale and proportional reasoning.

Having students free write about the visual prompts is ideal. You can make this a timed writing assignment where students must put writing implement to paper for a set amount of time, say 70 seconds. Beginning with a smaller

* The original color images can be downloaded in the web resources.

DOI: 10.4324/9781003374589-4

amount of time and increasing it over the semester or year will help students build stamina in free writing about a visual prompt. However, if some students need extra support, you can provide a prompt such as one of these:

- ❏ What do you see?
- ❏ How do you think math was used in this picture?
- ❏ What questions does the picture make you think about?
- ❏ What mathematical vocabulary could you use to describe the picture?
- ❏ Do you see any patterns in the picture? If so, describe the pattern.
- ❏ Where might you have seen something similar to what this picture is showing?
- ❏ How could you determine the height of the bear looking into the Denver Convention Center?
- ❏ What shapes do you see on the wall of prisms?
- ❏ If you drew a rectangle around the array of pipes on the truck, how much area of the rectangle would the circular ends of the pipes cover?
- ❏ How would you think about a speed limit of 8½ mph? What would you be riding in/on for this speed limit to make sense? Why?
- ❏ When you look at the steps in Highland, NC, why do you think there are so many landings? What role would slope play in the design of the steps? Explain.
- ❏ Discuss the path of the water shooting from the mouths of the frogs at the Dallas Arboretum.
- ❏ Discuss the grouping of Chihuly marbles being sure to include specific mathematical terms that could be used to describe the marbles.
- ❏ Research the works of Mexican artist Yvonne Domenge. How do you think she used mathematics when designing and making the Tabachin Ribbon?
- ❏ In the grouping of elevator photos, write a short story about what is found on each of the "unusual" floors numbered "1.5," "0," and "–1."
- ❏ Describe geometrically the Amber Fort located in Jaipur, India.
- ❏ Mathematics is often called the tool of the sciences. Study the science wall found in a high school. Choose at least three of the science topics represented. What mathematics might be used in each of the topics you choose? Be specific.

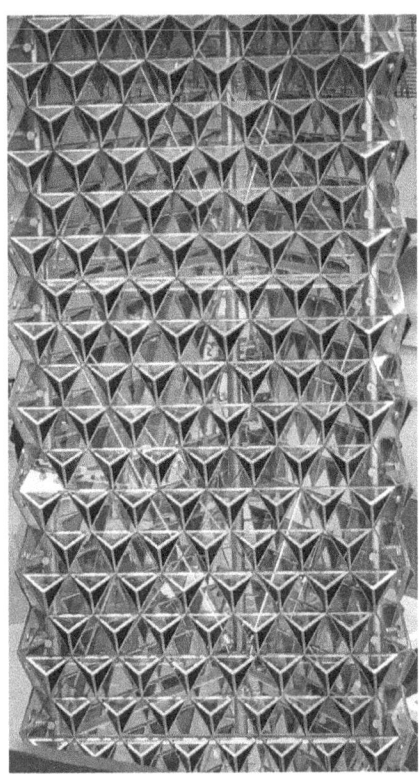

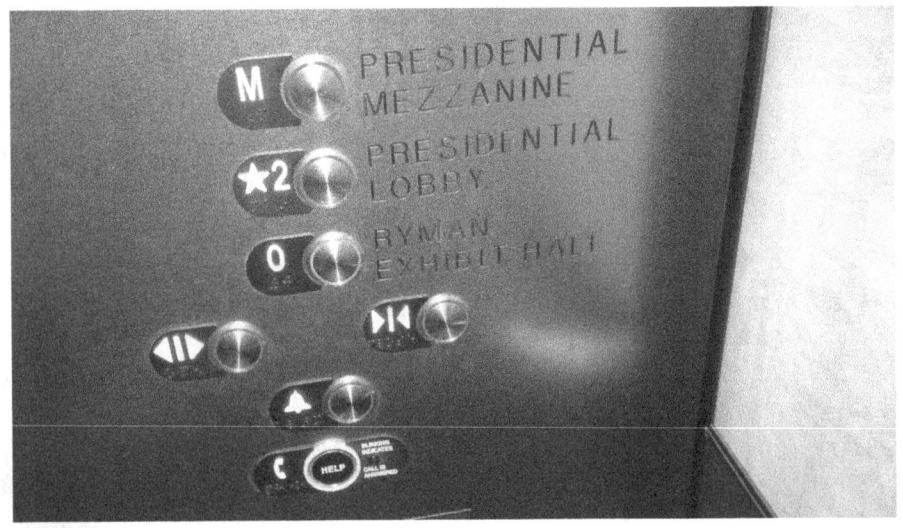

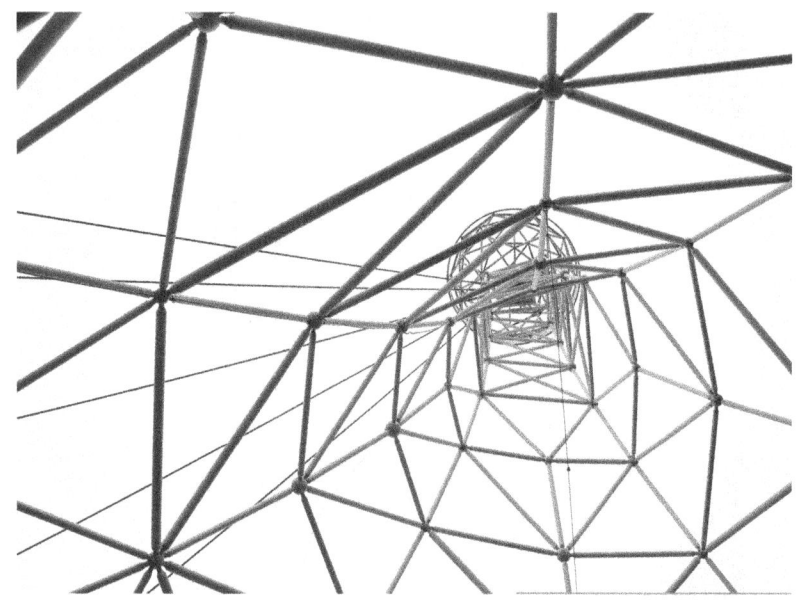

CHAPTER 3

Compare and Contrast

Writing math is typically a challenge for students. Using a Mathematician's Notebook "can change the way you teach as well as how your students learn and experience their content. The notebook becomes a dynamic place where language, data, and logical reasoning experiences operate jointly to form meaning for the student" (Jones & Texas, 2017). "A Mathematician's Notebook helps students create an organized space for demonstrating their learning process. The notebook serves as a formative instructional tool as well as a portfolio of the students' learning experiences." (Jones & Texas, p.14). Whether you are using a Mathematician's Notebook, an interactive notebook, or some other method of students chronicling their journey, all students need to be writing about math daily using paper and a writing implement.

Two of the main components of the Mathematician's Notebook are the glossary and the journal. Vocabulary is one of the foundations for developing understanding for any subject area and mathematics is no exception. Students need many opportunities to use their vocabulary in their daily work. Having students develop a glossary and reference the glossary as they progress through the year provides a resource for the students to use in their current mathematics course as well as future courses. Additional opportunities for students to engage with their academic vocabulary are vital for students to develop the deep understanding needed for success.

One such opportunity is the Compare and Contrast activity. Students can simply make a T-chart on their paper. They write the word pair (or three

DOI: 10.4324/9781003374589-5

columns if using three words), one word at the top of each column. Students then compare the words by listing the ways they are alike and different. They write their ideas in the columns below each word pair. They conclude by writing a summary sentence about their ideas. If time, students can complete additional pairs. There is a graphic organizer provided if desired to use. It is set up so when copied it can be cut in half and used with two students.

Interactive Word Walls and Strategy Walls

Ideally, the vocabulary used in this activity would already be displayed on a word wall of key terms that have been discussed throughout the instructional unit. A strategic way to make a word wall more interactive would be to use words from the wall for this activity. Assign students the words or allow student choice, which would reveal how students are making sense of the relationships between the concepts. Once the activity is complete, have students display their work on the wall alongside the words.

To reinforce the idea of students building a toolkit of strategies that can be used when problem solving, a strategy wall is a helpful anchor chart. Using words from the additional lists below (create a list, create a table, create a graph, draw a picture and draw a diagram, educated guess and random guess, eliminate possibilities and solve a simpler problem, formula and function, look for a pattern and use a formula, work backward and work forward, write an equation or inequality and model with manipulatives) build a strategy wall at the conclusion of the activity by displaying the words (problem-solving strategies) and student responses.

A Beginning List of Word Pairs

Topic 1: Number and Quantity
Absolute value of a number and sign of a number
Additive inverses and zero pairs
Constant of proportionality and slope
Decimals and fractions
Directly proportional and indirectly proportional
Distance on the number line and distance between two points
Fractions and rational numbers
Greatest Common Factor and Least Common Multiple
Integers and whole numbers
Percent and percent error
Positive, negative, and zero (this is intentionally vague as it could be related to integers, slope, the number line, etc.)

Proportional and not proportional
Rational numbers and irrational numbers
Ratios and fractions
Ratios and proportions
Unit fractions and unit rate
Unit rate and constant of proportionality

Topic 2: Algebraic Reasoning

Arithmetic and algebraic
Commutative property, associative property, and distributive property (commutative and associative can be for either addition or multiplication)
Dependent variable and independent variable
Domain and range
Equations and functions
Equations and inequalities
Evaluate and solve
Expressions and equations
Graphs and tables
Horizontal and vertical
Origin and ordered pair
Output and input
Radical and root
Solution and answer
Unit rate and slope
Variable and coefficient
Variable and quantity
x-axis and y-axis

Topic 3: Geometric Reasoning/Measurement and Units

Two-dimensional (2-D) and three-dimensional (3-D)
Area and volume
Compose and decompose (in terms of figures, both 2-D and 3-D)
Cone and cylinder
Congruence and similarity
Converse and conditional
Degrees, feet, and meters (as units of geometric measurements)
Dimensional analysis and unit conversions
Distance and length
Draw, sketch, and construct

Edges, faces, and vertices (you can also use apex, if studying pyramids/cones)

Linear, quadratic, and cubic (use this opportunity to connect geometric reasoning to algebraic reasoning)

Metric system and English standard system

Parallel lines, perpendicular lines, and skew lines

Perimeter, area, and volume

Perimeter and circumference

Pyramid and prism

Rectangular prism and hexagonal prism

Reflection and translation

Rotation and reflection

Scale, ratio, and unit rate

Scale factor and dilation

Supplementary angles and complementary angles

Translation and dilation

Vertical angles and adjacent angles

Topic 4: Data Analysis, Probability, and Statistics

Center, spread, and shape

Compound events and simple events

Conditional probability and basic probability

Counting principles and basic probability

Dot plot and scatterplot

Frequency and relative frequency

Histogram and box plot

Inference and observation

Interquartile range and mean absolute deviation

Linear association and nonlinear association

Making inferences and justifying conclusions

Mean, median, and mode

Measures of center and measures of variability

Positive association and negative association

Probability and chance

Range and spread

Single variable data and bivariate data

Spread and center

Theoretical probability and experimental probability

Tree diagrams, tables, and organized lists

Variability and spread

Variation and value

Additional Lists

Compare and convert

Compute and compose

Create a list, create a table, and create a graph

Draw a picture and draw a diagram

Educated guess and random guess

Eliminate possibilities and solve a simpler problem

Formula and function

Look for a pattern and use a formula

Markdown and markup

Work backward and work forward

Write an equation or inequality and model with manipulatives

Compare & Contrast

Choose a word pair. Write each word pair in
the boxes below. Compare the words by
listing the ways they are alike and different.
Write your ideas in the columns below each
word pair. Write a summary sentence about
your ideas.

Word pair	
Compare:	
Contrast:	

Summary sentence(s):

Compare & Contrast

Choose a word pair. Write each word pair in
the boxes below. Compare the words by
listing the ways they are alike and different.
Write your ideas in the columns below each
word pair. Write a summary sentence about
your ideas.

Word pair	
Compare:	
Contrast:	

Summary sentence(s):

CHAPTER 4

The Answer Is...

Students benefit from open-ended questions where there is possibly more than one correct response. This writing strategy allows students the opportunity to think beyond just procedural solving to get "the" answer. In some cases, the context is set up and given for the students. These questions will offer you as well as your students insight into how they think about mathematics. Open-ended questions also encourage growth mindset.

Students choose a card from "The Answer Is..." set to write about. Or you can assign one based upon students' individual needs. Students read the setup, if one is provided, then, they create a contextual problem for which the solution would be the answer given. This writing activity can be easily differentiated by setting parameters for students. The contextual problem can be single step, or it may be multistep. It could require a specific operation or include quantities within specific parameters. Students could also be required to provide at least two different possibilities for a context where the solution is the answer. Drawings, illustrations, and labels might also be needed for a complete response.

DOI: 10.4324/9781003374589-6

Topic 1: Number and Quantity	
You are working with a rectangle. The answer you get is $\sqrt{2}$ What could the question be?	You are working with a change in temperature. The answer you get is −4°F. What could the question be?
The answer you get is $3\frac{2}{5}$ of a candy bar. What could the question be?	You are buying produce. The answer is $0.85 per pound. What could the question be?

Topic 2: Algebraic Reasoning	
You are combining algebraic expressions. The answer you get is $$2.5x - \frac{1}{3}.$$ What could the question be?	You have been working with distances and rates. The answer you get is 43 mph.  What could the question be?
You are working with inequalities. The answer you get is $$\{-3,-2,-1,0,1\}.$$ What could the question be?	You are working with functions in context. You get the resulting graph shown below. 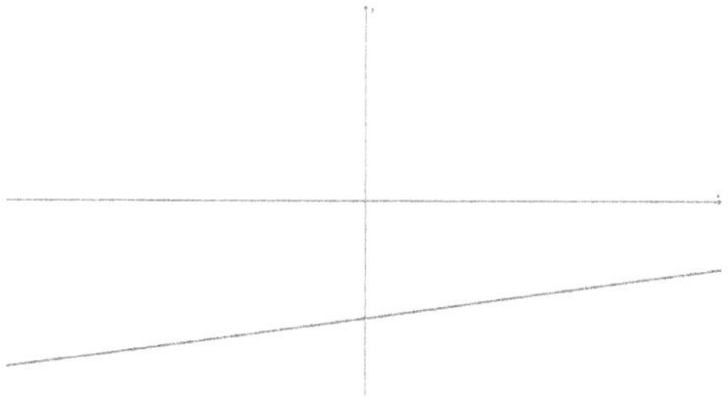 What could the question be? Be sure to include a context.

Topic 3: Geometric Reasoning/Measurement and Units	
You are working in the coordinate plane. The answer is Quadrant 3. 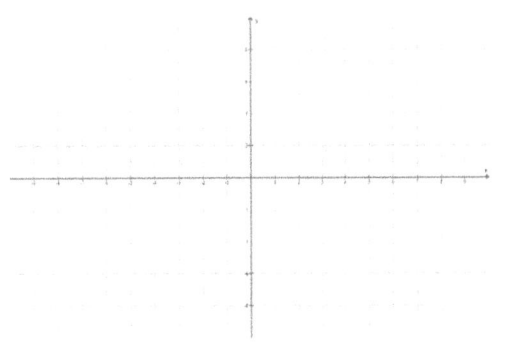 What could the question be?	You have been working at a bakery making birthday cakes. The answer you get is 7π cm.  What could the question be?
You are working with nets and three-dimensional (3-D) figures. The answer you get is at least one square.  What could the 3-D figure be? Draw figure and net. What could the question be?	You were asked to draw a triangle similar to the one given. Your answer is the triangle below. 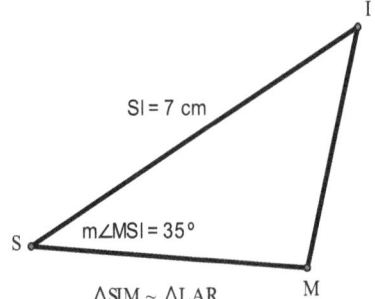 SI = 7 cm m∠MSI = 35° S △SIM ~ △LAR M What could the original triangle have looked like? Include angle and side measures. Angle M is an obtuse angle.

You have been working with composing and decomposing shapes and perimeter and area. The answer you get is at least 30 square feet.

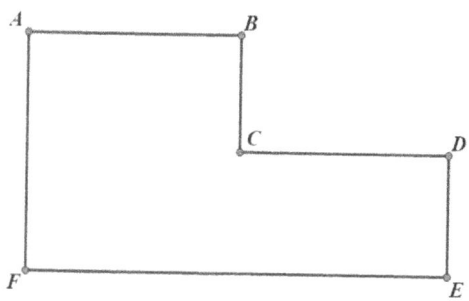

What could the dimensions be? What could the question be?

You have been working with volumes and surface areas of 3-D figures. The final calculation for the volume of a 3-D figure is shown below.

$$V = 1200 \text{ m}^3$$

What could the 3-D figure be? What could be its dimensions?

Topic 4: Data Analysis, Probability, and Statistics	
You have been working with determining measures of center. The answer you get is the mean is 74.5. $$\bar{x} = 74.5$$ What could the question be? Include the data set and context.	You are working with scatterplots. The answer is a negative linear correlation. What could the question be? Create the context and the related scatterplot.
You have been working with basic probabilities with coins, number cubes, and cards. The answer you get is, "The probability is $\frac{1}{3}$". What could the question be?	

CHAPTER 5

Topical Questions

Questions are tools in a teachers' toolbox and should be used as chisels to promote student thinking rather than pliers for answer-getting. Teachers practice, refine, and hone their questioning skills as they engage with students daily. Students can only provide a depth of answer based upon the quality of the question(s) asked.

As discussed in Chapter 1, there are four opportunities for questioning students that naturally exist when students are working through a task or activity. And the questions need to be carefully crafted so the mathematical discourse is not shut down. Using a question stem such as "is," "do," or "could" allows students the opportunity to simply answer yes or no and then they are done. If, however, you use a stem such as "how," "what," "when," and "where," along with others listed on our Q-Pyramid overlay (Jones & Texas, p. 92), you have opened the conversation and students must engage in mathematical discourse. This use of a more inquiring form of response encourages students to justify or explain their responses, whether they be correct or incorrect.

As you hone your questioning technique, be aware that one of the most important parts of questioning is how you respond to students. You should respond to your students in a manner that supports their thinking as it reveals to you what and how they are thinking. Wait time is vital as a quick response can often shut down the individual's or rest of the class's thinking and/or reflection on what is being said. Asking students to explain why or to further discuss

DOI: 10.4324/9781003374589-7

how they thought about something may at first be a struggle with students, but if it becomes a consistent part of your questioning, students will eventually accept it.

The questions provided in this section are not universal, but rather nuanced to the topic they reference. By design, they require communicating the answer more fully and are perfect for encouraging students to write about math.

Topic 1: Number and Quantity

General

- ❏ What is the meaning of percent?
- ❏ How does comparing quantities describe the relationship between the quantities?
- ❏ When is it more beneficial to use percentages? Decimals? Ratios? Why?

Subtopic Specific

- ❏ **Rational numbers**

    - ❏ How can fractions, decimals, and percents represent the same rational number?
    - ❏ How do rational numbers relate to whole numbers?
    - ❏ How do multiplicative inverses relate to the identity element for multiplication?
    - ❏ How does operating with integers compare to operating with rational numbers?
    - ❏ What is the relationship between a repeating fraction and a decimal?
    - ❏ Are all ratios rational numbers? Why or why not?
    - ❏ Are all fractions ratios? Why or why not?

- ❏ **Integers**

    - ❏ What is the relationship between "the opposite of," "negative," and "subtract?" Give a context to support your thinking.
    - ❏ How do integers relate to whole numbers?
    - ❏ What is meant by the term "zero pair"?
    - ❏ How can "zero pairs" aid in solving equations?
    - ❏ How do additive inverses relate to the identity element for addition?

- ❏ **Irrational numbers**

    - ❏ How do irrational numbers relate to integers?
    - ❏ How can you determine if a decimal cannot be written as a rational number?
    - ❏ Why is pi considered an irrational number? What is the difference in using an approximation for pi when calculating the area of a circle from using the exact irrational number?

❏ David says that all numbers under a radical sign are irrational. Claudine disagrees. Who is correct and why? (Note – neither – both are depending upon...).

❏ How can radicals help determine a length or distance on the coordinate plane?

Topic 2: Algebraic Reasoning

General

- ❏ Why would you use multiple representations of linear equations and inequalities?
- ❏ When do evaluate versus when do you solve when working with algebraic reasoning?
- ❏ How can expressions be extended to create equations and/or functions? Give a specific example.

Subtopic Specific

- ❏ **Expressions**
    - ❏ What does it mean to have an equivalent expression?
    - ❏ How can you determine if an algebraic algorithm is both effective and efficient?
    - ❏ What is a variable?

- ❏ **Equations**
    - ❏ What does it mean to be a solution to an equation?
    - ❏ What does it mean to be a solution to a system of equations?
    - ❏ How does the use of equivalent expressions aid when solving equations?
    - ❏ What evidence is needed to determine how many solutions exist for an equation? For a system of equations?
    - ❏ What are the benefits and limitations of solving a system of equations using graphing, substitution, and elimination? Give an example of a system that would best be solved by each of the methods.
    - ❏ What does "=" mean? Be very specific with your response. (equal vs same vs answer)

- ❏ **Inequalities**
    - ❏ What does it mean to be a solution to an inequality?
    - ❏ What does it mean "to be true" when discussing the/a solution(s) to an equation? An inequality?
    - ❏ How can constraints or conditions in a real-world context be represented by inequalities? Explain using a specific example.
    - ❏ Which properties of equality do not hold for inequalities? Why? Show specific examples?

❑ **Functions**

    ❑ What is the difference between a relation (that is not a function) and a function?

    ❑ Can a function's input value share an output value with a different input value? Explain.

    ❑ How are patterns of change related to the behavior of functions?

    ❑ How can patterns, relations, and functions be used as tools to best describe and help explain real-life situations?

    ❑ What is the difference between a function and an equation?

Topic 3: Geometric Reasoning/ Measurement and Units

General

- ❑ How can geometry help us make sense of our world?
- ❑ What is the difference in a sketch, a drawing, and a construction in geometry?
- ❑ What is a dimension?
- ❑ How do measuring and labeling units help us make sense of our world? Be specific.
- ❑ Why can different units represent the same measurement?
- ❑ How can tables of equivalent ratios, tape diagrams, double number line diagrams, and equations help solve contextual situations?

Subtopic Specific

- ❑ **Coordinate Plane**

  - ❑ Who is credited with the development of the coordinate plane and for what purpose was it created?
  - ❑ Why do you think the quadrants are numbered in the order they are? Support your reasoning. (Students need to make connections to the domain, the x-axis and the independent variables, including both positive and negative values.)
  - ❑ How can distance be described using the coordinate plane?
  - ❑ It can be said that the coordinate plane is a tool of visualization. What contextual situations support this idea? Explain.
  - ❑ How does the convention for naming points on the coordinate plane support algebraic reasoning? (Give a nod to why understanding this helps students not make common mistakes later, and why discussions around the historical connections help students better understand other topics and may prevent future misconceptions.)

- ❑ **Circles**

  - ❑ Lexi says a circle is a polygon. Rosie disagrees. Who is correct and why?
  - ❑ What is the minimum needed to define a circle?
  - ❑ How are pi and a circle related?
  - ❑ Tammy bragged that if you gave her just one measure of a circle, she could determine all other measurements for that circle. How could you prove or disprove Tammy's statement?

❑ **Nets**

   ❑ Nets are one of the most useful representations in geometry. How do nets help create objects like soda cans, cereal boxes, and backpacks? Be specific.

   ❑ How do nets provide a bridge from two-dimensional (2-D) to three-dimensional (3-D) objects?

   ❑ What are two examples of a cross section/slice of a 3-D object that would result in a triangle? In a rectangle? In a circle? An oval/ellipse? (Students should include the geometric figure that describes the object as well as how it was sliced...for example, if students use a cone, when the slice and the ice cream are parallel to the opening, they get a circle; if the slice is at an angle, it is an oval/ellipse.)

   ❑ How does decomposing a 3-D figure into a net help determine the surface area for the figure?

❑ **Scale drawings**

   ❑ How do scale drawings and the actual measurements model a proportional relationship?

   ❑ If you double a side length for an object, by what scale factor is the perimeter changed, the area, and/or the volume? Explain.

   ❑ What would be an example of a context for which the scale of the drawing to the scale of the object was 1:0.01? Explain.

❑ **Congruence and Similarity**

   ❑ How do you show that two triangles are similar based on their angles alone?

   ❑ How do compositions of transformations model the commutative property, the associative property, and/or the distributive property?

   ❑ What is the difference between similar figures and congruent figures?

   ❑ Can it be said that all circles are similar? Why or why not? (Are they polygons?)

   ❑ When is a figure, which is transformed in the plane, similar to the original figure?

   ❑ How is congruence related to translation?

   ❑ How can you determine whether figures are congruent through your reasoning about rigid transformations?

   ❑ If a side of one triangle is congruent to a side of another triangle, what information about their angles would allow you to prove the triangles congruent?

   ❑ How do you identify the corresponding parts of congruent triangles?

❏ **Area**

❏ What is the relationship between the area of a circle and the area of a parallelogram?

❏ How does decomposing a complex figure into other shapes help you determine the figure's area?

❏ How are perimeter, area, and volume related to algebraic functions? (perimeter, linear; area, quadratic; volume, cubic)

❏ How can figures have the same area but different perimeters? Different shapes?

❏ Under what conditions would a parallelogram, a triangle, and a square all have the same area? Be specific? (Note: how to differentiate from just sketching with basic labels to showing that each of the figures can be decomposed into two congruent triangles – think tangrams.)

❏ **Volume**

❏ Eson says that if he knows how to determine the area of a figure, he does not need to memorize volume formulas. When is this possible? When is this not possible?

❏ How does calculating the volume of prisms compare to calculating the volume of cones, pyramids, and spheres? Explain.

❏ How can figures have the same volume but different surface areas? The same surface area but different volumes?

❏ **Scales**

❏ How are scales and units related?

❏ When creating a scale drawing, how would the units affect the drawing?

❏ How can scales be used to misrepresent data?

❏ **Dimensional Analysis**

❏ Why does dimensional analysis allow one not to have to memorize formulas? What do you have to have memorized or know for this to be true?

❏ When is dimensional analysis a more efficient conversion technique? When is it not?

Topic 4: Data Analysis, Probability, and Statistics

General

- ❏ How can you tell that a graph contains a mistake or is intentionally misleading? (Hint: You cannot tell by just looking at the graph.)
- ❏ How does statistics help you understand the world?
- ❏ What are some ways that data is being collected on you every day?
- ❏ What are the advantages and disadvantages of analyzing data by hand versus using technology?

Subtopic Specific

- ❏ **Variability/Distributions**
  - ❏ How does the term "variable" in statistics compare with how the term variable is used in algebraic reasoning?
  - ❏ What does it mean for data to be distributed?
  - ❏ How would you compare the variability of two different distributions?
  - ❏ What are examples of situations where you would want to compare two different distributions?
  - ❏ How can each of the three centers for a set of data be visualized on a dot plot?
  - ❏ Patrick says that a graph that is skewed right looks backward from how "skewed right" is defined. Jamal disagrees. With whom do you agree? Why? Support your reasoning.

- ❏ **Inferences**
  - ❏ What is the difference between prediction and inference?
  - ❏ How can random sampling be used to draw inferences about a population?
  - ❏ How does the degree of visual overlap in two numerical data distributions with similar variabilities lead to making a comparative inference about the two populations?

- ❏ **Single Variable Data**
  - ❏ How does a variable in statistics compare to a variable in algebra? (In algebra, a variable is a unique quantity where in statistics a variable, such as "r" (range), describes a sample of a population or the population itself.)

❏ What are some ways that two data sets can have the same range but be different?

❏ How does the context of the problem inform the type of center you choose to use?

❏ Will data with a larger range always have a higher standard deviation?

❏ What would a standard deviation of zero mean?

❏ How do the different graphs emphasize different aspects of the data? What is an outlier? How might you identify one?

❏ **Bivariate Data**

  ❏ What is bivariate data and how do two-way tables involve bivariate data?

  ❏ How could you visualize a two-way frequency table graphically?

  ❏ How can relative frequency tables be used to summarize data?

  ❏ Does the order in which you plot points on a scatterplot matter? What does each point/dot represent on the scatterplot? Explain.

  ❏ Do you think perfect positive correlation and negative correlation exist often between variables? Explain.

❏ **Probability**

  ❏ How are probability and chance related?

  ❏ What is the difference between theoretical and experimental probabilities?

  ❏ What is the difference between independent and dependent events?

  ❏ How can a sample space be represented with a list?

  ❏ How is representing a sample space different than finding the possible outcomes?

  ❏ How does an outcome table relate to the tree diagram it represents? Be specific.

CHAPTER 6

Writing About...

Writing About is a small group writing activity that can be used strategically to support students who struggle with writing, particularly language learners. Just because a student can verbally tell you something does not mean that they can write that same response and support it with evidence. Prior to this activity you might invite the English Language Arts (ELA) teacher to visit the class and share what makes a good paragraph so common expectations can be set that support the work in ELA.

Begin by giving students two or three index cards or scraps of paper. Students are to study the word cloud and write one or two sentences about the topic using words they find in the word cloud. Each student shares their sentences with the group and together create a paragraph about the topic. The index cards allow students to sequence the sentences to build a thoughtful and complete paragraph. They combine similar sentences and check for an introduction, conclusion, etc. This provides an opportunity for students to practice building a paragraph about a topic. As students first work in a group of three or four, they can then begin to work with a smaller group or a partner. The activity can be extended later as an individual writing activity as students are developing stamina for writing. Be aware that not all students will progress at the same pace.

Extensions: Students can sort the words found in the word cloud and create a mapping. Students can work in small groups, pairs, or individually. Students need to be able to articulate their sorting/mapping rule. If students are

DOI: 10.4324/9781003374589-8

doing a mapping, they can draw connectors, use string/yarn, or use something like WikkiStix™. If using WikkiStix™, be sure students are working on a piece of construction paper or scrap paper that will not matter if the sticky gets on it or not.

Once students sort their word set and show their connections, they need to write down their sorting rule in their Mathematician's Notebook. Once all groups are finished, students can do a Walk About Review where they observe the other groups' mapping/sorting and make notes about what they think their sorting rules were. Then, the whole group can come back together and discuss what they observed. Some questions that you might use to facilitate the discussion could include:

- ❏ What were the similarities you observed between the mappings?
- ❏ What were some differences?
- ❏ Were you able to identify the correct sorting rule for the other groups? Why or why not?

Suggested directions for the mapping/sort:

Study the words. Sort the words. Sort the sets of words that seem to go together. You may use your string/WikkiStix™ to show connections between the words. Explain your sorting rule fully. If directed, create a second sorting with a different rule.

Suggested directions for the Walk About Review:

As you walk about and review the other groups' mappings, do not talk, look over the mapping, and in your notebook identify what you think the groups' sorting/mapping rule is and why. You will have a set amount of time at each mapping, so use it wisely and efficiently.

Ideas for Display: Groups can create a graffiti board using a chart paper to capture their paragraph. These group boards can then be put together to create a graffiti wall. The class could do a gallery walk to view what was developed, provide feedback, and/or reflect on the process.

Writing about...

Study the word cloud below. Create at least two statements about data analysis using the key words you see in the word cloud. With your group, use your sentences to create a paragraph about data analysis.

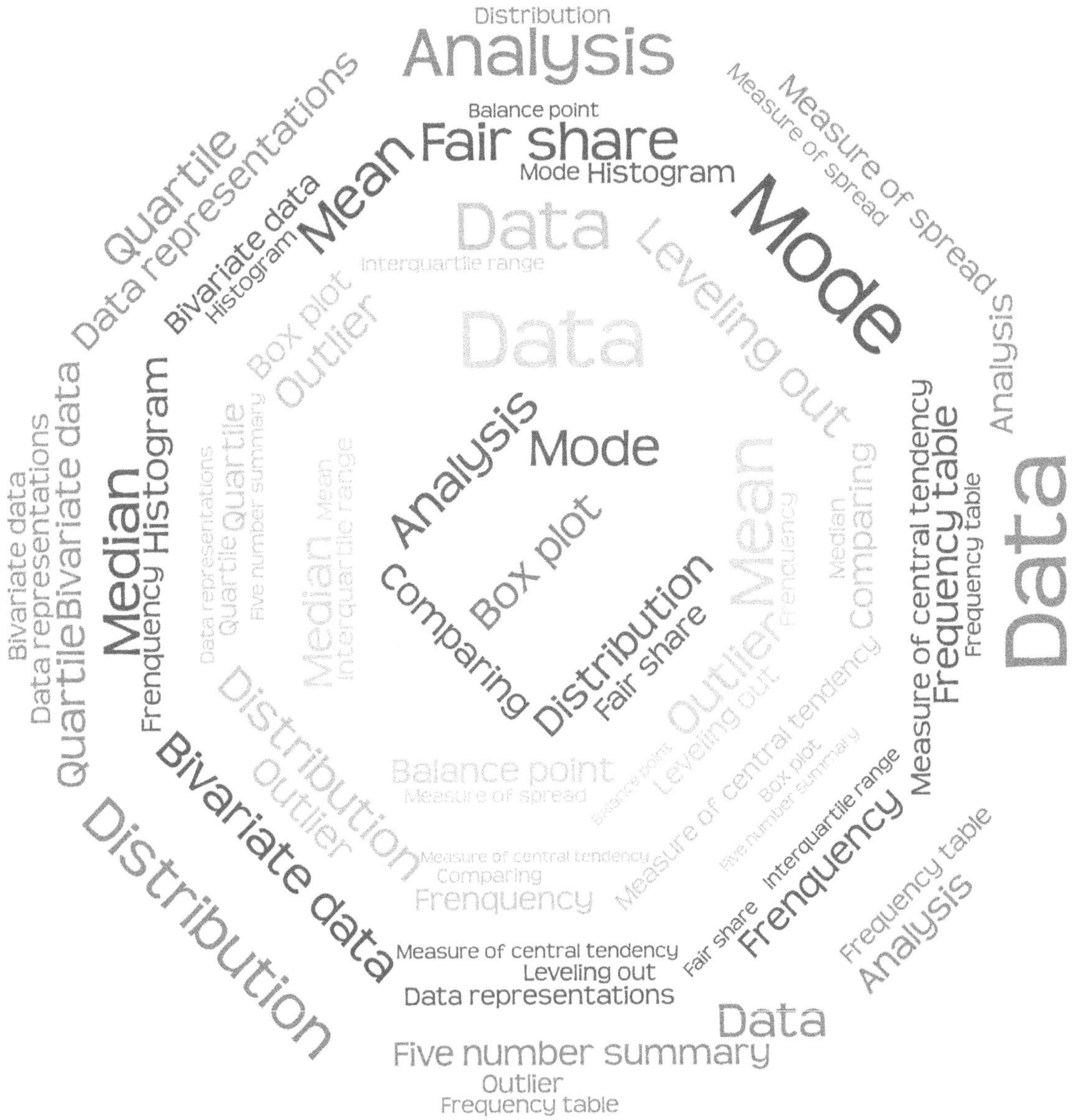

Writing about...

Study the word cloud below. Create at least two statements about expressions and equations using the key words you see in the word cloud. With your group, use your sentences to create a paragraph about expressions and equations.

Writing about...

Study the word cloud below. Create at least two statements about geometry using the key words you see in the word cloud. With your group, use your sentences to create a paragraph about geometry.

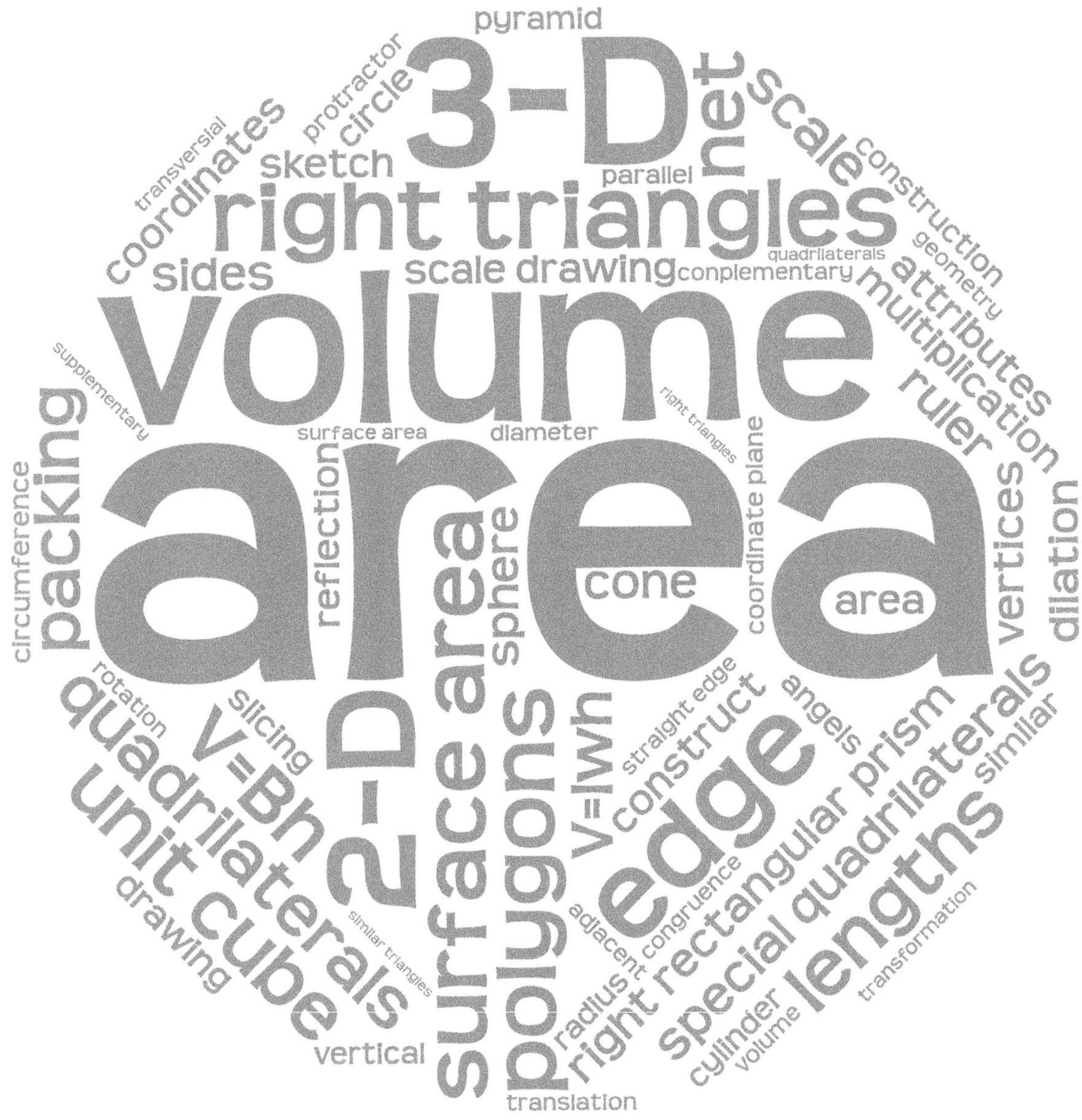

pyramid
transversial protractor circle **3-D** net scale
coordinates sketch parallel construction
right triangles scale drawing complementary attributes geometry
sides quadrilaterals multiplication
volume
supplementary surface area diameter right triangles ruler
circumference packing reflection **area** vertices dilation
2-D surface area sphere cone area
rotation slicing polygons straight edge construct angels right rectangular prism similar
quadrilaterals V=Bh edge special quadrilaterals lengths
unit cube similar triangles V=lwh adjacent congruence cylinder transformation
drawing radius volume
vertical translation

Writing about...

Study the word cloud below. Create at least two statements about inequalities using the key words you see in the word cloud. With your group, use your sentences to create a paragraph about inequalities.

Writing about...

Study the word cloud below. Create at least two statements about linear functions using the key words you see in the word cloud. With your group, use your sentences to create a paragraph about linear functions.

Writing about...

Study the word cloud below. Create at least two statements about probability using the key words you see in the word cloud. With your group, use your sentences to create a paragraph about probability.

Sample
Estimate
Inference
Population proportion
Distribution
Distribution
Probability
Tree diagrams
Sample
Theoretical
Simulation
Variability
Event
Experimental
Dependent events
Lists sample mean
Experimental Theoretical
Event
Data
Event Sample
Likely
independent events
Sample
Tables Experimental
Event Simulation
Actual probability
Population proportion
chance
Random
Probability mode
Relative frequency
Law of large numbers
Relative frequency
Sample mean
Sample numbers
Median
simulation
Law of large numbers
Inference
Estimate
Probability diagram
Chance
unlikely
compound event

Writing about...

Study the word cloud below. Create at least two statements about ratios and proportional reasoning using the key words you see in the word cloud. With your group, use your sentences to create a paragraph about ratios and proportional reasoning.

Rate
coordinate axes
representations
double number line
Quantity
slope triangles
proportional relationship
constant of proportionality
measurement unit conversion
rate of change
ratio table
unit pricing
graph
tape-diagram
proportion
Variation
fraction
unit rate
equation
percent
Relationship
comparing ratios
ordered pairs
constant speed
Ratio
slope
variables

Writing about...

Study the word cloud below. Create at least two statements about the real number system using the key words you see in the word cloud. With your group, use your sentences to create a paragraph about the real number system.

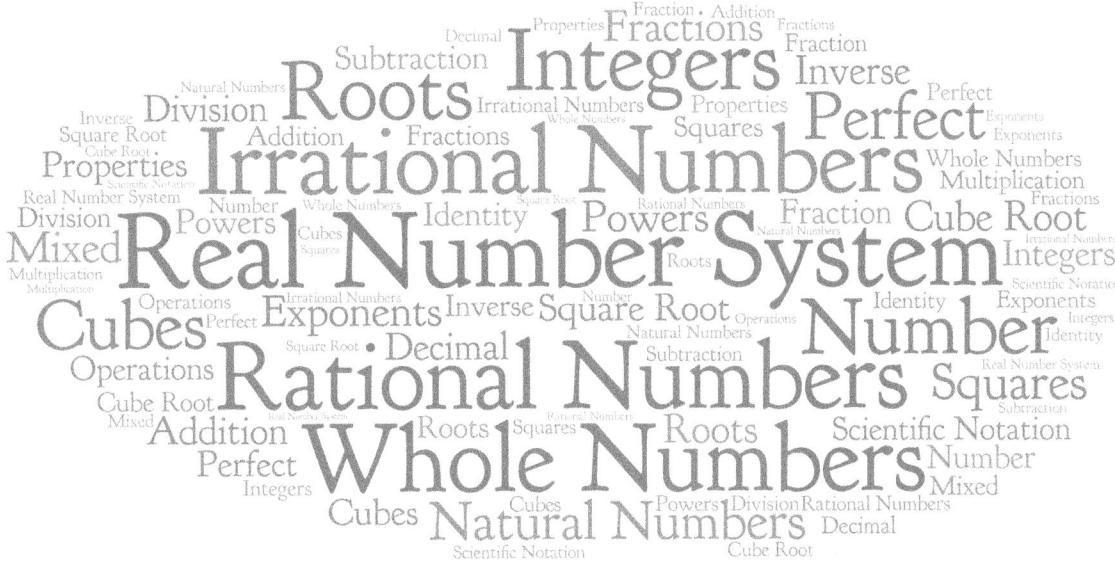

Topic 1: Number and Quantity

Addition	**Cubes**
Cube Root	**Decimal**
Division	**Exponents**

Fractions	**Identity**
Integers	**Inverse**
Irrational numbers	**Mixed numbers**

Multiplication	**Natural numbers**
Number	**Operations**
Perfect	**Powers**

Properties	Rational numbers
Real number system	**Roots**
Scientific notation	**Square root**

Squares	**Subtraction**
Whole numbers	

Comparing ratios	**Constant of proportionality**
Constant speed	**Coordinate axes**
Double number line	**Equation**

Fraction	**Graph**
Measurement unit conversion	**Ordered pairs**
Percent	**Proportion**

Proportional reasoning	**Quantity**
Rate of change	**Rate**
Ratio table	**Ratio**

Relationship	**Representations**
Slope triangles	**Slope**
Tape diagram	**Unit pricing**

Unit rate	Variables
Variation	

Topic 2: Algebraic Reasoning

Algebraic	**Analyze**
Answers	**Arithmetic**
Calculate	**Change**

Coefficient	**Collecting like terms**
Compare	**Computation**
Constant	**Constraint**

Context	**Cube**
Decimal	**Determine**
Distance	**Distributive property**

Equation	Equivalent
Estimate	Evaluate
Expand	Exponent

Expression	**Factor**
Formula	**Graph**
Identity	**Independent**

Inequality	Infinite
Interpret	Intersection
Linear	List

Multistep	**Nonlinear**
Notation	**Number**
Operations	**Parentheses**

Perfect square	**Positive**
Power	**Product**
Property	**Proportional**

Quotient	**Radicals**
Root	**Set**
Simultaneous	**Slope**

Solution	Solve
Speed	Square
Substitution	Symbols

Technology	**Term**
Transform	**Unit**
Unknown	**Value**

Variable	**Vertical**
Write	

Circle	**Condition**
Constraint	**Dot**
Equal to	**Equality**

False	Graph
Greater than or equal to	Greater than
Inequality	Infinite

Less than	**Less than or equal to**
Number line	**Set**
Solution	**Solving**

Substitution	True
Values	

Change	**Constant rate**
Context	**Decreasing**
Dependent	**Domain**

Equal differences	**Equation**
Evaluate	**Function**
Graph	**Grow**

Increasing	**Independent**
Intercepts	**Linear function**
Model	**Negative**

Notation	Per unit
Positive	Range
Relation	Representation

Rise	**Run**
Slope	**Table**
Undefined	**Verbal**

x-Axis	y-Axis
Zero	

Topic 3: Geometric Reasoning/ Measurement and Units

2-D	3-D
Adjacent angles	**Area**
Attributes	**Circle**

Circumference	**Cone**
Congruence	**Complementary angles**
Construction	**Coordinate plane**

Coordinates	**Cylinder**
Diameter	**Dilation**
Drawing	**Edge**

Geometry	Net
Packing	Parallel
Perpendicular	Polygons

Protractor	**Pyramid**
Quadrilaterals	**Radius**
Reflection	**Right triangular prism**

Right triangle	**Rotation**
Ruler	**Scale drawing**
Sides	**Similar triangles**

Sketch	**Slicing**
Sphere	**Straight edge**
Supplementary angles	**Surface area**

Transformation	**Translation**
Transversal	**Unit cube**
V = Bh	**V = lwh**

Vertical angles	**Vertices**
Volume	

Topic 4: Data, Analysis, Probability, and Statistics

Bivariate data	**Box plot**
Data	**Data analysis**
Data representations	**Distribution**

Five number summary	**Frequency**
Frequency table	**Histogram**
Interquartile range	**Leveling out**

Mean	**Measures of central tendency**
Measure of spread	**Median**
Mode	**Outlier**

Quartile	**Range**
Skewness	

Chance	Compound events
Data	Dependent events
Discrepancy	Distribution

Estimation	**Event**
Experimental probability	**Independent events**
Inference	**Law of large numbers**

Likely	**Lists**
Odds	**Population**
Population proportion	**Prediction**

Random	**Relative frequency**
Sample	**Sample mean**
Simulation	**Tables**

Theoretical probability	Tree diagrams
Unlikely	Valid
Variability	

Journal Prompts

As mentioned previously in the Compare and Contrast section, two of the main components of the Mathematician's Notebook are the Glossary and the Journal. Journals are a great way for students to keep track of their mathematical journey as well as giving you insight into how they think about math and how they have developed and grown over the course of the semester or year. Journals can be a place where students engage with quotes, historical connections to topics, famous people related to the topic of study, misconceptions, and 'What if?' scenarios. If you keep a 'parking lot' in your classroom for students to list issues they had with the homework from the day before, if only a couple of people had trouble with, let's say problem #2, we might just have other students work that at the board and then see what follow-up questions, if any, were needed. But if ten students had issues with problem #2, then as the teacher, I must ask, "What did I not do that allowed students access to beginning the problem." So, that problem might go into the journal with a discussion and notes around issues students were having.

Journal entries are best assessed separately from the rest of the Mathematician's Notebook. Even a basic *All, Most, Some, None* format works well. Journals are a place that you can dialogue with students about topics and engage with them in a different way. The authors never wrote directly on the student's notebook pages, as that was their own work, but wrote on sticky notes and attached it to the page(s) where comments were appropriate.

DOI: 10.4324/9781003374589-9

Following are some beginning suggestions for journal prompts that can be used throughout the year. Some are very focused around mathematical topics, and some are just for fun for students to allow their imagination to run wild. Hopefully, these will serve as the basis for you to add many additional ideas of your own.

Math Specific

Math-ography: Students write about their prior experiences with mathematics. No comments on specific teachers allowed. What topics made sense, what topics were challenging, and how do they see mathematics fitting into their occupational plans.

WRITE YOUR "MATH-OGRAPHY"

Include your:

- Earliest remembrances of counting and learning about numbers
- Elementary school work and topics you 'got' as well as topics that were challenging
- Your goals for this class
- What you see yourself doing after you complete high school and what role math will play in that...

DO NOT name teacher names!

A self-evaluation: Usually assigned around the first interim period for grading, this provides the opportunity for students to reflect on their work so far. It has proved helpful to give students a list of questions to guide their reflections. Some examples are:

- ❏ Do you have a dedicated place at home to study?
- ❏ Have you been regular in your attendance?
- ❏ How would you rate your engagement in class when you are here?
- ❏ How much time outside of class are you spending on work for this class?
- ❏ Do you feel that you are getting enough sleep and rest?
- ❏ How are your eating habits affecting your schoolwork?

Writing to Explain

Option 1: Students write an explanation for a student in their class who was absent the day they learned about/how to (insert topic/activity/procedure for the day here).

Option 2: Students are assigned a mathematical topic such as one of the geometric transformations or measures of central tendency. Then they complete the following:

- ❏ You are a rotation. Tell us everything we should know about you.
- ❏ You are the mode of a set of data. Tell us everything we should know about you.

Creative Writing

Students complete each of the following prompts using their imagination. Encourage students to just not write but also use drawings and sketches and color as they complete the prompt.

- ❏ If I were a number, I would be _____ because....
- ❏ If I were a geometric shape, I would be _____ because....
- ❏ You wake up tomorrow morning and find that circles no longer exist. How do you get to school? Be sure to sketch the road, vehicle, etc., that you would use. Extension: How else will circles not existing impact your life? Be specific.
- ❏ If I were a mathematical pattern, I would be _____ because....
- ❏ What I find the most challenging with _____ (current topic) is.... Explain why.
- ❏ When I see a math problem with words, I feel _____ because....
- ❏ Choose a character from literature and describe how you might use mathematics in what you do.

Using Quotes

Option 1: Students write the quote, they write what it meant in the time it was written, and then how it would be applicable to them in math class today.

Example: I apologize for the length of this letter. Had I but more time it would have been shorter. – Blaise Pascal

- ❏ Students copy the quote in their journal.
- ❏ They then write a couple of sentences about what they think Pascal meant in light of his time period. They may need to ask their history

or English teacher for help here. Back in Pascal's time, people wrote drafts of their letters, much as students write drafts for their English papers today. So, he did not have a lot of time to "clean up" his letter and make it shorter.

❏ Students conclude by writing a couple of sentences about how this will apply to their work in math class. When writing in mathematics, in their Mathematician's Notebook, etc., students need to learn to not only be precise but be succinct with their explanations and to the point.

Option 2: Students write a response to the author.

Option 3: Students write about what questions the quote prompts them to think about.

Option 4: Students describe what the quote means to them.

Below is a beginning list of quotes which span from historical to modern day, includes a diverse group of individuals, and cuts across disciplines to include the humanities as well as the sciences.

I am not what I think I am, I am not what you think I am, I am what I think you think I am. – Charles Cooley[1]

It's amazing what one can do when one doesn't know what one can't do. – Garfield the Cat[2]

I have made this letter longer than usual, because I lack the time to make it short. – Blaise Pascal[3]

Number is the within of all things. – Pythagoras[4]

What we know is not much. What we do not know is immense. – Pierre-Simon Laplace[5]

Nature's Great Book is written in mathematical symbols. – Galileo Galilei[5]

If you don't like the answer, ask a different question. – Dr. Larry Fleinhardt (NUMB3RS)[5]

One accurate measurement is worth a thousand expert opinions. – Rear Admiral Grace Hopper[5]

The purpose of models is not to fit the data but to sharpen the questions. – Samuel Karlin[5]

A thing is obvious mathematically after you see it. – R.D. Carmichael[5]

The different branches of Arithmetic – Ambition, Distraction, Uglification, and Derision. (from Alice in Wonderland*)* – Lewis Carroll[6]

[The universe] cannot be read until we have learnt the language and become familiar with the characters in which it is written. It is written in mathematical language, and the letters are triangles, circles and other geometrical figures, without which means it is humanly impossible to comprehend a single word. – Galileo Galilei[7]

It is impossible to be a mathematician without being a poet in soul. – Sofia Vasilyevna Kovalevskaya[8]

Formulation of a problem is often more essential than its solution which may be merely a matter of mathematical or experimental skill. – Alfred North Whitehead[4]

I have finally found a subject where I do not need to memorize, but can think things out myself: mathematics. – Herta Freitag (from her diary, age 12)[5]

Any fool can know. The point is to understand. – Albert Einstein[8]

What is mathematics? It is only a systematic effort of solving puzzles posed by nature. – Shakuntala Devi[9]

There was virtually no aspect of twentieth-century defense technology that had not been touched by the hands and minds of female mathematicians. – Margot Lee Shetterly[10]

Mathematics is the most beautiful and most powerful creation of the human spirit. – Stefan Banach[9]

Mathematics knows no races or geographic boundaries; for mathematics, the cultural world is one country. – David Hilbert[9]

There should be no such thing as boring mathematics. – Edsger W. Dijkstra[9]

'Obvious' is the most dangerous word in mathematics. – Eric Temple Bell[9]

Math proficiency is the gateway to a number of incredible careers that students may never have considered. – Danica McKellar[11]

Don't let anyone rob you of your imagination, your creativity, or your curiosity. It's your place in the world; it's your life. Go on and do all you can with it, and make it the life you want to live. – Mae Jemison[12]

Simple laws can very well describe complex structures. The miracle is not the complexity of our world, but the simplicity of the equations describing that complexity. – Sander Bais[13]

Division is esteemed one of the busiest operations of Arithmetic, and such as requireth a mind not wandering, or settled upon other matters. – Thomas Hylles[13]

Idealism increases in direct proportion to one's distance from the problem. – John Galsworthy[11]

Life without geometry is pointless. – Unknown[8]

We learn more by looking for the answer to a question and not finding it than we do from learning the answer itself. – Lloyd Alexander[8]

You don't understand anything until you learn it more than one way. – Marvin Minsky[8]

The most useful piece of learning for the uses of life is to unlearn what is untrue. – Antisthenes[8]

Knowledge is like money: to be of value it must circulate, and in circulating it can increase in quantity and, hopefully, in value. – Louis L'Amour[8]

A lot of music is mathematics. It's balance. – Mel Brooks[11]

Euclid's first common notion is this: Things which are equal to the same things are equal to each other. That's a rule of mathematical reasoning and its true because it works. – Abraham Lincoln[14]

References

1. https://www.goodreads.com/author/quotes/5953688.Charles_Horton_Cooley
2. https://www.gocomics.com/garfield/1982/01/19
3. https://mathshistory.st-andrews.ac.uk/Biographies/Pascal/quotations/
4. https://quotefancy.com/
5. https://mathshistory.st-andrews.ac.uk
6. https://www.britannica.com/quotes/Lewis-Carroll
7. https://www.mathnasium.com/blog/why-mathematics-is-a-language
8. https://www.livingmath.net/quotes
9. https://www.prodigygame.com/main-en/blog/math-quotes/
10. https://www.goodreads.com/work/quotes/45855800-hidden-figures-the-american-dream-and-the-untold-story-of-the-black-wom
11. https://www.brainyquote.com/quotes/
12. https://yourstory.com/herstory/2019/11/inspirational-quotes-women-stemm-sheryl-sandberg
13. https://www.famousscientists.org/magnificent-mathematics-quotes/
14. https://ficquotes.com/quotes/4926/

CHAPTER 8

Poetry/Prose

In the spirit of writing in response to a quote (described in the Journal Prompts), this section begins with one from JoAnne Growney's blog Intersections – Poetry with Mathematics:

> Mathematical language can heighten the imagery of a poem; mathematical structure can deepen its effect.

The precision of language required of both disciplines makes the intersection of mathematics and poetry seem almost obvious. Providing the opportunity to see the connection allows students to explore this relationship while deepening their understanding of mathematics and writing skills.

Acrostic: explain – one word, expression, describing the main word...

Example:

Circles

Centers
Inside
Radii and
Chords
Lopsided never
Elegant with perfect
Symmetry

DOI: 10.4324/9781003374589-10

Math

Mysterious to some
Ability required
The language of our world
Hurts my brain!

Beginning List of Terms (See Word Lists in Compare and Contrast (Chapter 3) for Additional Terms)

Equation

Expression
Geometry (or any specific shape or term)
Inequality
Math
Numbers (or use any specific number set)
Probability
Proportional
Ratio
Statistics

Fibonacci poem: Students can create their own "Fibonacci" poem where each line of the poem has the number of words as found in the sequence. This can be differentiated for students by having them use only the first three or four numbers found in the sequence or more if their writing skills allow. The topic of the poem can be of their choosing or can be assigned. The following poem models 1, 1, 2, 3, 5, 8.

Triangles

Triangles
Pointed
Three sides
And three angles
Walking from vertex to vertex
Around the corners, either obtuse, acute, or right.

Haiku: It has three lines, five syllables in the first and third lines, and seven syllables in the second line.

Example:

Ratios

Ratios compare
Proportional reasoning
Ratios equal

Pi poem: It is similar to the Fibonacci poem. Students write lines based on the digits in pi: 3.1415926535 8979323846 2643383279. This can be differentiated for students by having them use only the first three or four digits found in pi or more if their writing skills allow.

Students can also have fun with the topic as in the example below.
Example:

Pi

My favorite pie –
cherry.
Flakey crust, tart flavor,
Yum!

Free verse/free write: There is no specific form, meter, or rhyme scheme. Students can have the freedom to write as they feel. Some suggestions are given below for types of free writes students may enjoy doing.

Cartoons

Commercial, Infographic, Public Service Announcement (PSA)
Free Verse Poem
Graphic novel, for example, *The Adventures of Slope Boy*
Historical Fiction
Math Carols and/or seasonal songs
Math words to a current song
Short Story

CHAPTER 9

Cubing and Think Dots

Cubing and think dots are two strategies for differentiation in the classroom. Traditionally, students are given a cube with a variety of activities or tasks at a given level. Different cubes can contain different levels of tasks and activities. Think dots work in a similar way. Cards with a certain number of dots are provided as well as a number cube. Students roll the number cube and work through the associated activity or task on the card with the corresponding number of dots. Again, tasks and activities are varied or leveled to meet the needs of the students. You can choose the parameters for your students or create your own set of think dot cards using index cards and practice problems from your chosen curriculum.

As students are working through the various activities, emphasize that they are just not to "show their work" but to also make statements explaining and supporting their reasoning and thinking. Students need to be able to use precise mathematical language and symbols in their written work as well as clearly articulate their thinking.

Preparation

Print the cubes and think dot cards on cardstock or heavy paper. Build the cubes carefully using the dotted lines as the folds. Tape or glue the edges

DOI: 10.4324/9781003374589-11

together. Fold the think dot cards and tape together. These resources can be used in a variety of ways in a center or learning station as well.

Materials List

- ❏ Set of think dot cards
- ❏ Action cube
- ❏ Constraint cube
- ❏ Number cubes and/or dominoes (Note: You can purchase domino sets that show numbers greater than six.)

Expressions

In this adaptation of a cubing and think dots activity, there are two cubes. One cube has action to perform on expressions. Another cube provides constraints for the actions as students work with expressions to evaluate them. Students will also use one or two number cubes, or they can use dominoes to create the fractions, integer values, etc. There are a variety of activities that students can engage in on the set of the six think dot cards.

Action Cube Options

Evaluate: Students use the action cube first. They roll the action cube, and then they roll a number cube to see which think dot card they use. If students roll "Evaluate," they roll the constraint cube to determine the type of number they use. They use a number cube or domino to determine the quantity they use. If a fraction is required, they can roll the number cube twice, once for the numerator and once for the denominator. Alternatively, they can pick a domino and use one side for the numerator and one for the denominator.

Combine: Students roll the action cube first. They then roll a number cube to see which think dot card they use. If students roll "Combine," they roll a number cube to see which of the bullets on the think dot card they combine. Students roll the number cube at least twice, but this can be adjusted based on your individual student's needs.

Simplify: Students roll the action cube first. They then roll a number cube to see which think dot card they use. If students roll "Simplify," they simplify each of the expressions on the chosen think dot card. If Card Four is chosen, students must first transform the expressions into their equivalent symbolic form. Then, simplify.

Identify Like Terms: Students roll the action cube first. They then roll a number cube to see which think dot card they use. If students roll "Identify Like Terms," they do so for each of the expressions on the chosen think dot card.

Identify Parts of the Expressions: Students roll the action cube first. They then roll a number cube to see which think dot card they use. If students roll "Identify Parts of the Expressions," they do so for each of the expressions on the chosen think dot card.

Write the Verbal Form/Symbolic Form: Students roll the action cube first. They then roll a number cube to see which think dot card they use. If students roll "Write the Verbal Form/Symbolic Form," they write the form for each of the expressions on the chosen think dot card that is not given. For example, if the verbal form is given, students write the algebraic/symbolic form and vice versa.

Geometric Reasoning

In this adaptation of a cubing and think dots activity, there are only three cubes and then several sets of geometric figure cards. One cube has an action to perform on geometric figures. Another cube provides constraints for the actions as students work with two-dimensional (2-D) figures. A third cube provides constraints for the actions as students work with three-dimensional (3-D) figures. It is your choice which constraint cube you have students work with or if you want them to work with both. There is a blank set of think dot cards you can print out and assign different practice probes from the curriculum you are using, if desired.

Action Cube Options

Compose/Decompose: Students use the action cube first. If students roll "Compose/Decompose," they roll the constraint cube to determine the type of shape they use. This can be differentiated based upon the needs of your students and where they are in their studies. They choose one of the geometric image cards to determine the figure they use to decompose or the figures they use to compose a shape.

Draw a Net: Students roll the action cube first. If students roll "Draw a net," they roll the 3-D constraint cube to see which of the shapes they draw a net of. Alternately, they can pick a geometric 3-D image card to draw a net of a more complex composite shape.

Seek and Find: Students roll the action cube first. If students roll "Seek and Find," they roll the constraint cube to determine the type of shape they use. This can be differentiated based upon the needs of your students and where they are in their studies. They then go on a scavenger hunt to find at least three examples of the shape they are using and document the item and sketch it to show how it models their assigned shape.

Create a Geometric Shape: Students roll the action cube first. If students roll "Create a Geometric Shape," they use the geometric drawing cards determine if a shape with the given parameters can be created. If they can create it, they draw the given shape with the given parameters, labeling the shape as needed. If the shape is unique, they should state that. If more than one drawing could be made given the parameters, encourage students to make a second drawing. If it is not possible, students state that the drawing is not possible as well as giving justification for their answer. You can choose which cards and how many cards your students use based upon their individual needs.

Create a Scale Drawing: Students roll the action cube first. If students roll "Create a Scale Drawing," they then select a scale drawing card and create a scale drawing for the item listed on the card. Cards can be assigned to meet the individual needs of your students. Be sure that you have examples of each of the items listed on the cards available in your classroom and/or school. Or you can assign items that you have available.

Calculate: Students roll the action cube first. If students roll "Calculate," they calculate the perimeter, area, surface, or volume for each of the figures on the chosen geometric figure card as you assign. You can choose if students work with 2-D figures or 3-D figures based on their individual needs as well as what measurement attribute of the figure they calculate. Students measure the figures on the cards using metric or English standard as you assign based on their need to practice with decimals and/or fractions. You can also assign specific practice problems from the curriculum you are using.

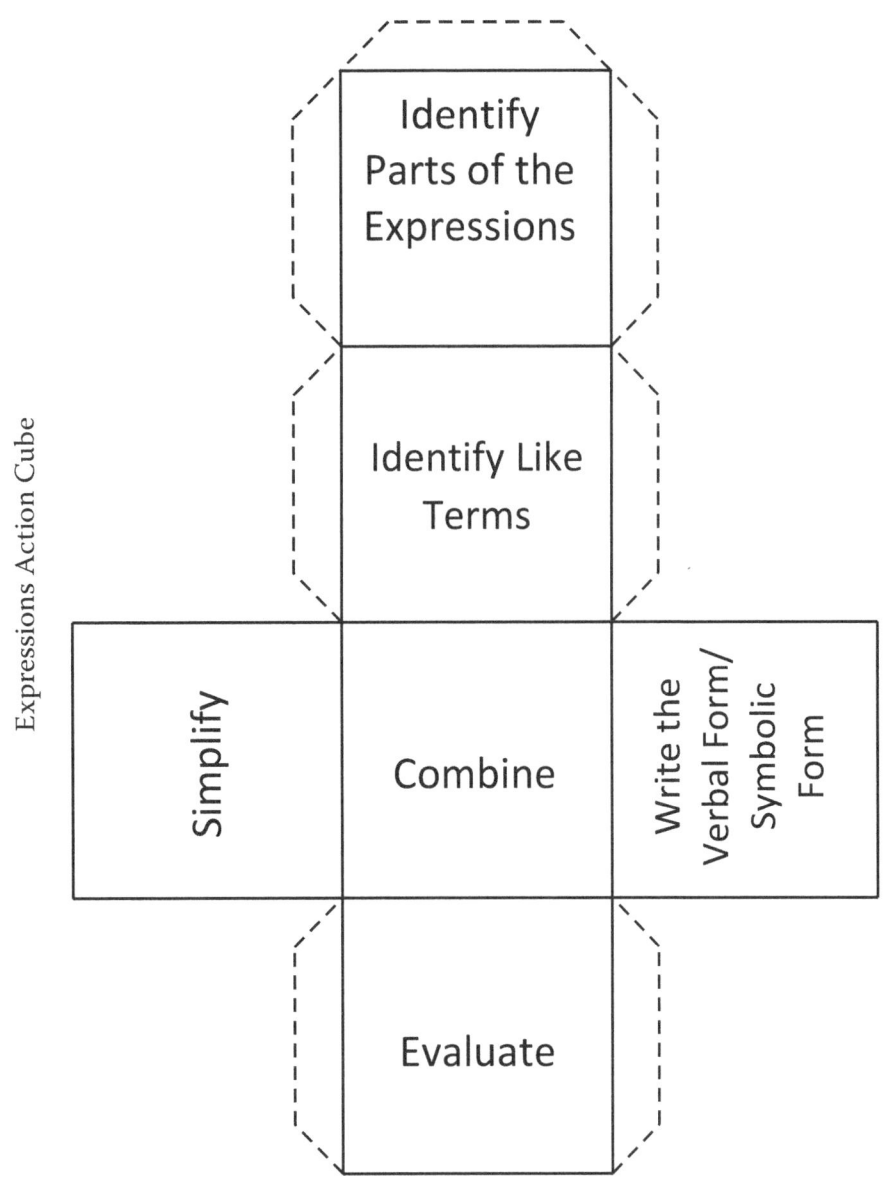

Expressions Action Cube

Identify Parts of the Expressions

Identify Like Terms

Simplify

Combine

Write the Verbal Form/ Symbolic Form

Evaluate

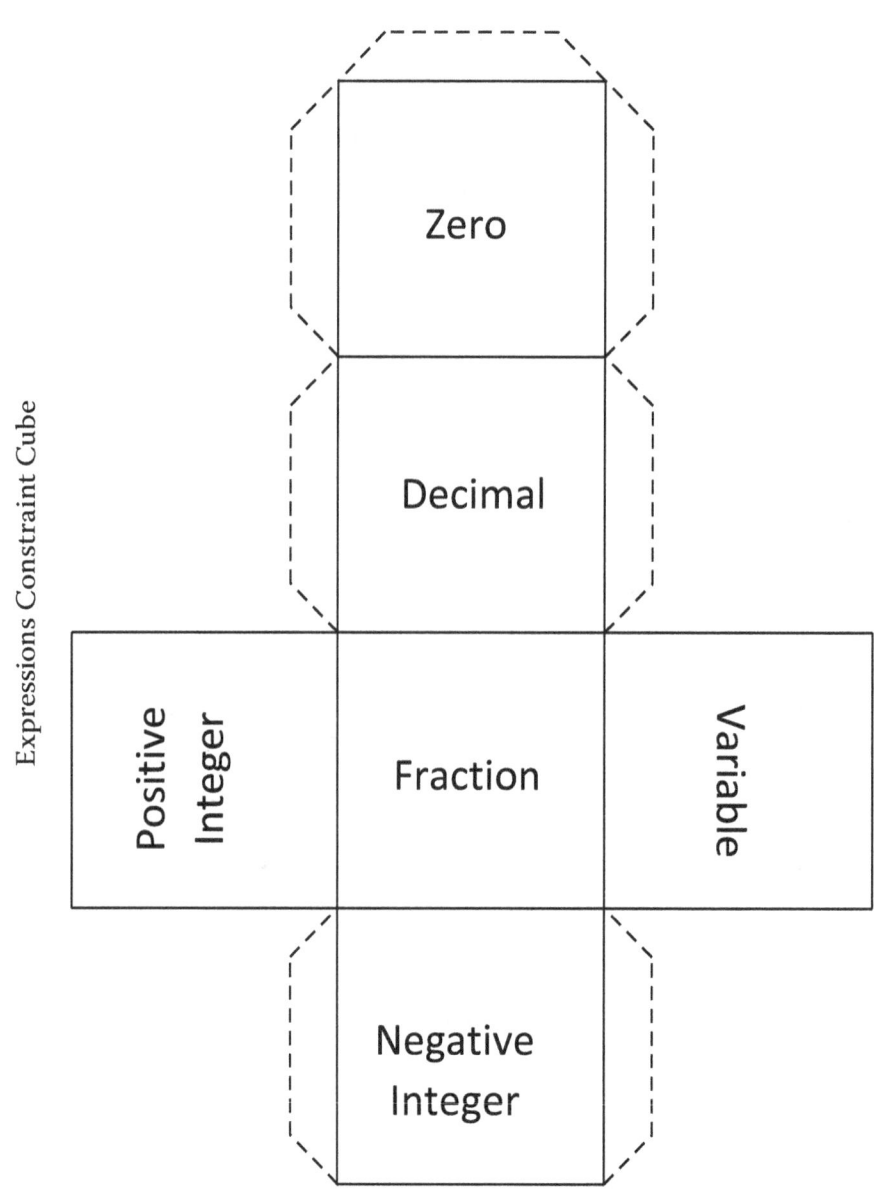

Expressions Constraint Cube

Think Dot Cards

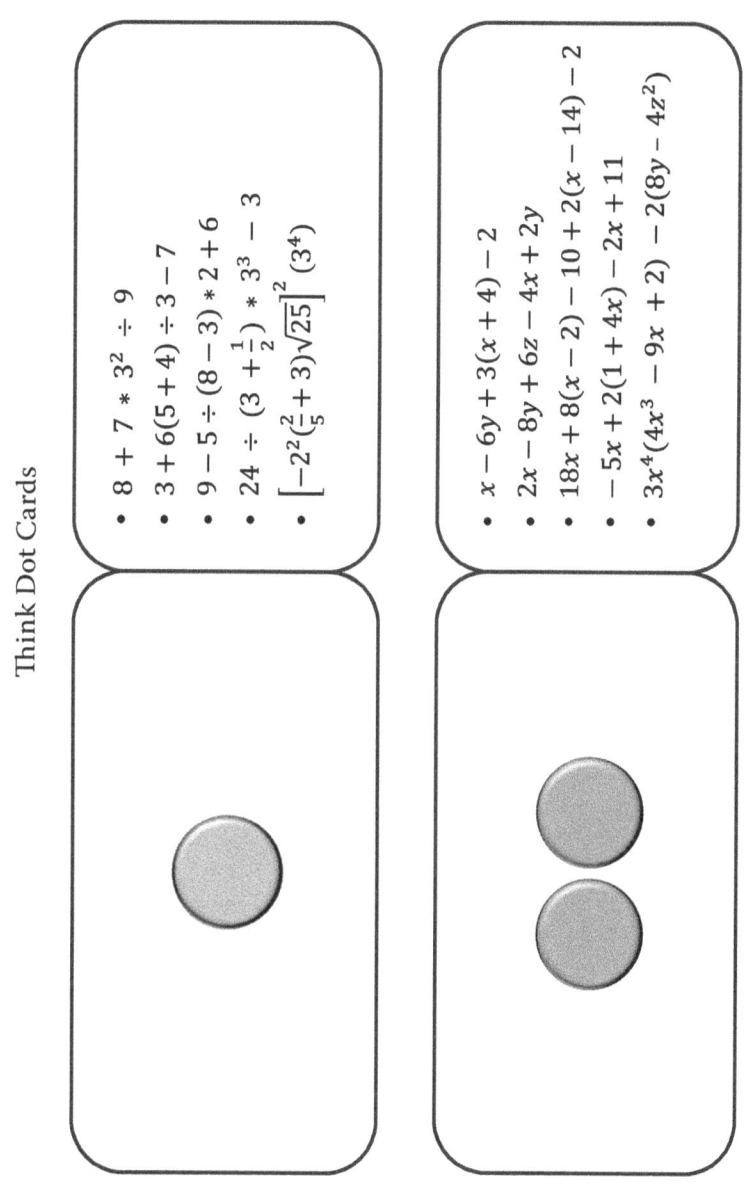

- $8 + 7 * 3^2 \div 9$
- $3 + 6(5 + 4) \div 3 - 7$
- $9 - 5 \div (8 - 3) * 2 + 6$
- $24 \div (3 + \frac{1}{2}) * 3^3 - 3$
- $\left[-2^2(\frac{2}{5} + 3)\sqrt{25}\right]^2 (3^4)$

- $x - 6y + 3(x + 4) - 2$
- $2x - 8y + 6z - 4x + 2y$
- $18x + 8(x - 2) - 10 + 2(x - 14) - 2$
- $-5x + 2(1 + 4x) - 2x + 11$
- $3x^4(4x^3 - 9x + 2) - 2(8y - 4z^2)$

Expressions

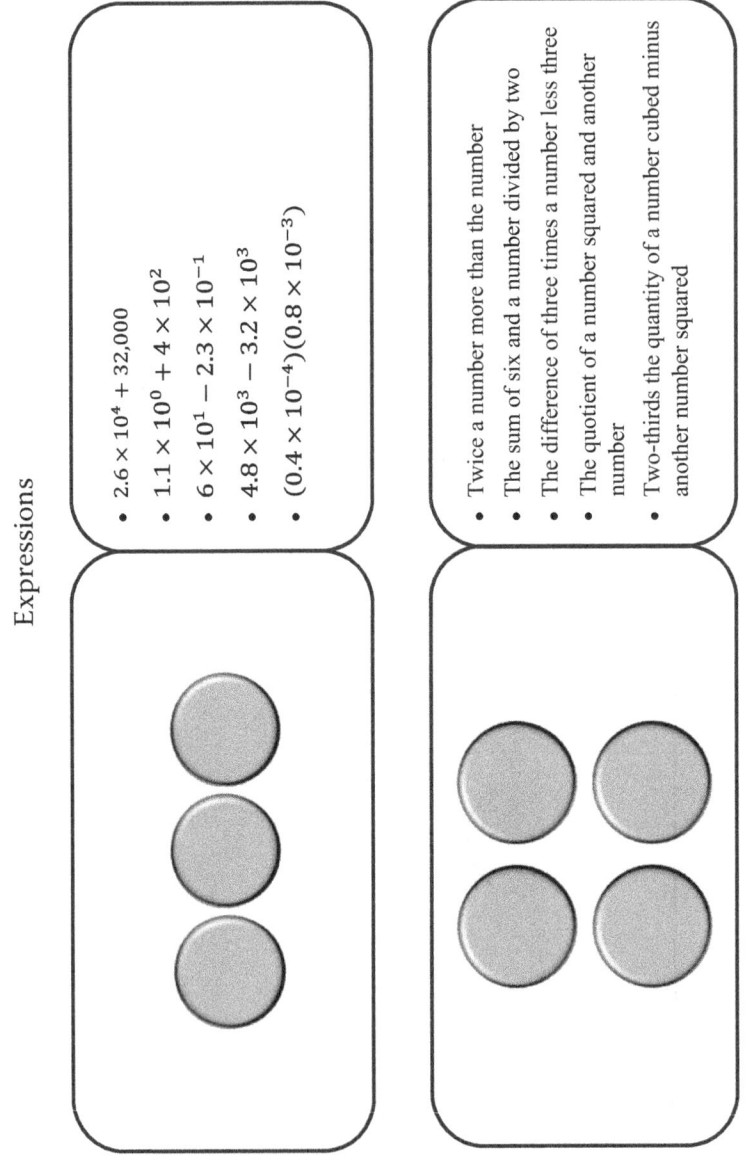

- $2.6 \times 10^4 + 32{,}000$
- $1.1 \times 10^0 + 4 \times 10^2$
- $6 \times 10^1 - 2.3 \times 10^{-1}$
- $4.8 \times 10^3 - 3.2 \times 10^3$
- $(0.4 \times 10^{-4})(0.8 \times 10^{-3})$

- Twice a number more than the number
- The sum of six and a number divided by two
- The difference of three times a number less three
- The quotient of a number squared and another number
- Two-thirds the quantity of a number cubed minus another number squared

Geometry Action Cube

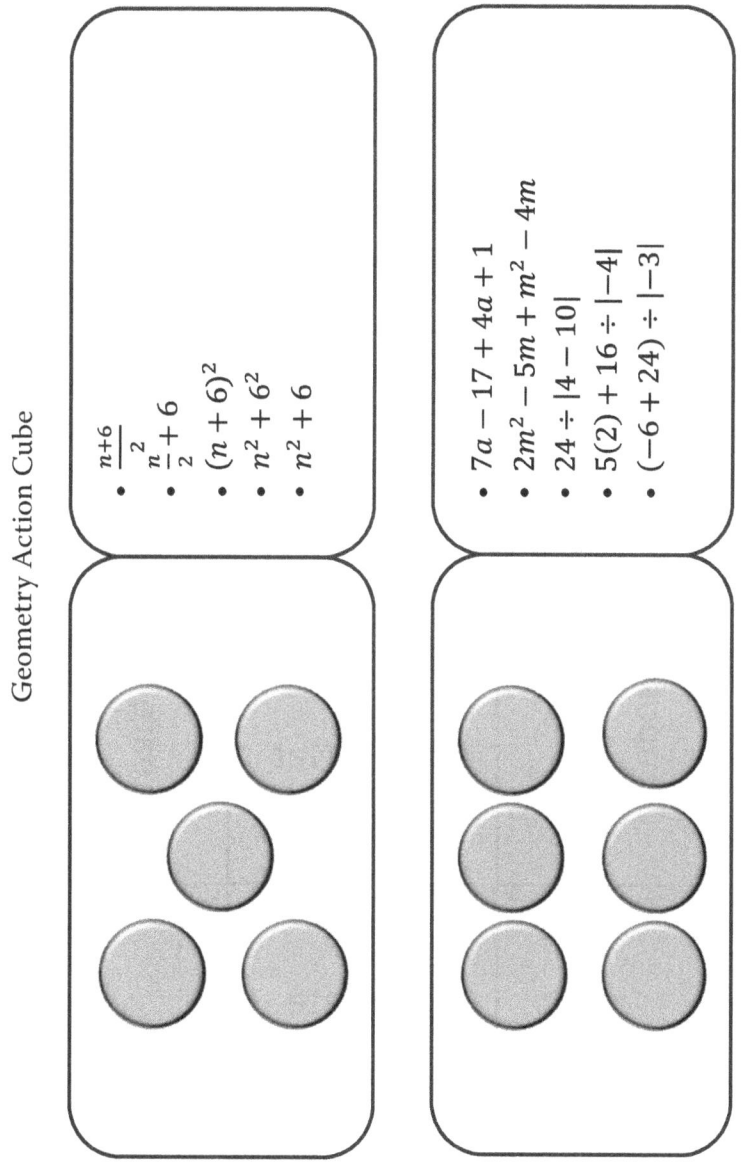

- $\dfrac{n+6}{2}$
- $\dfrac{n^2}{2} + 6$
- $(n+6)^2$
- $n^2 + 6^2$
- $n^2 + 6$

- $7a - 17 + 4a + 1$
- $2m^2 - 5m + m^2 - 4m$
- $24 \div |4 - 10|$
- $5(2) + 16 \div |-4|$
- $(-6 + 24) \div |-3|$

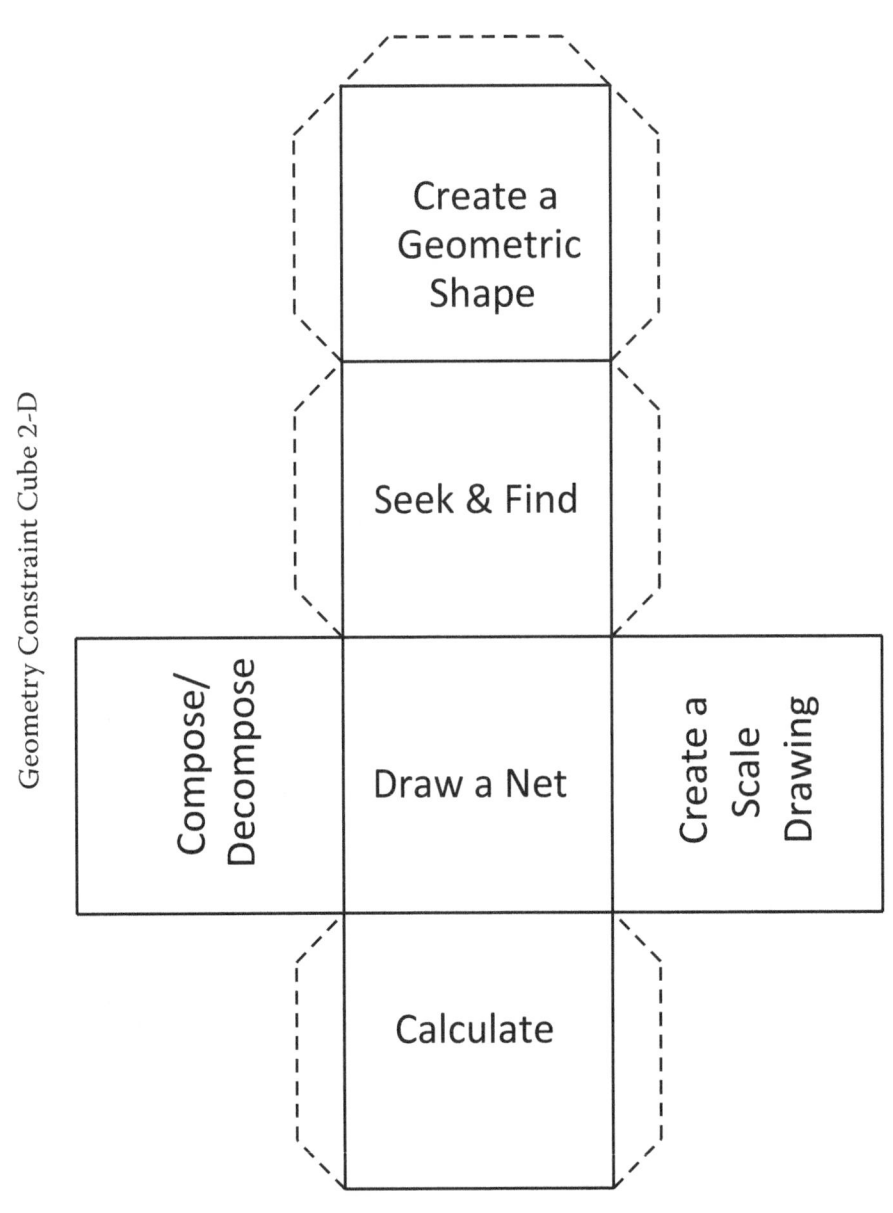

Geometry Constraint Cube 2-D

Create a
Geometric
Shape

Seek & Find

Compose/
Decompose

Draw a Net

Create a
Scale
Drawing

Calculate

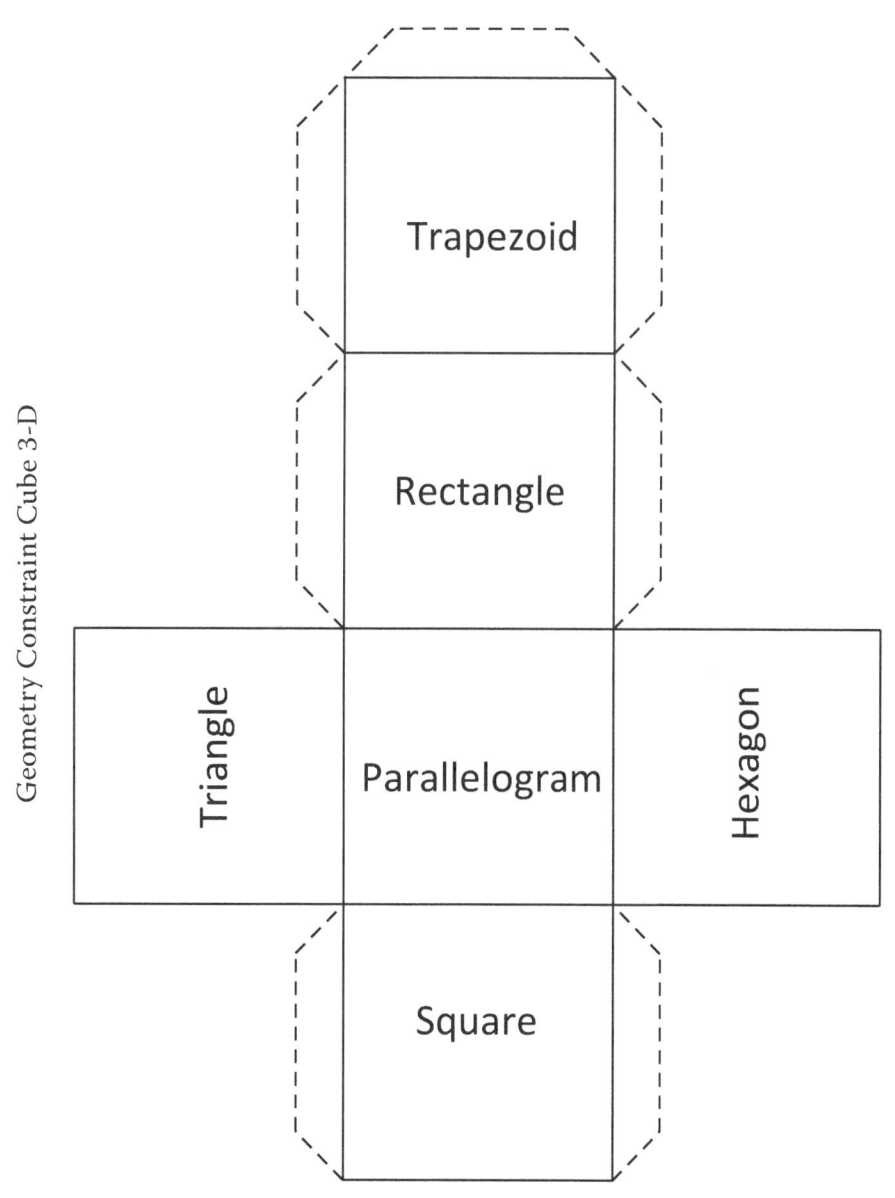

Geometry Constraint Cube 3-D

Trapezoid

Rectangle

Triangle

Parallelogram

Hexagon

Square

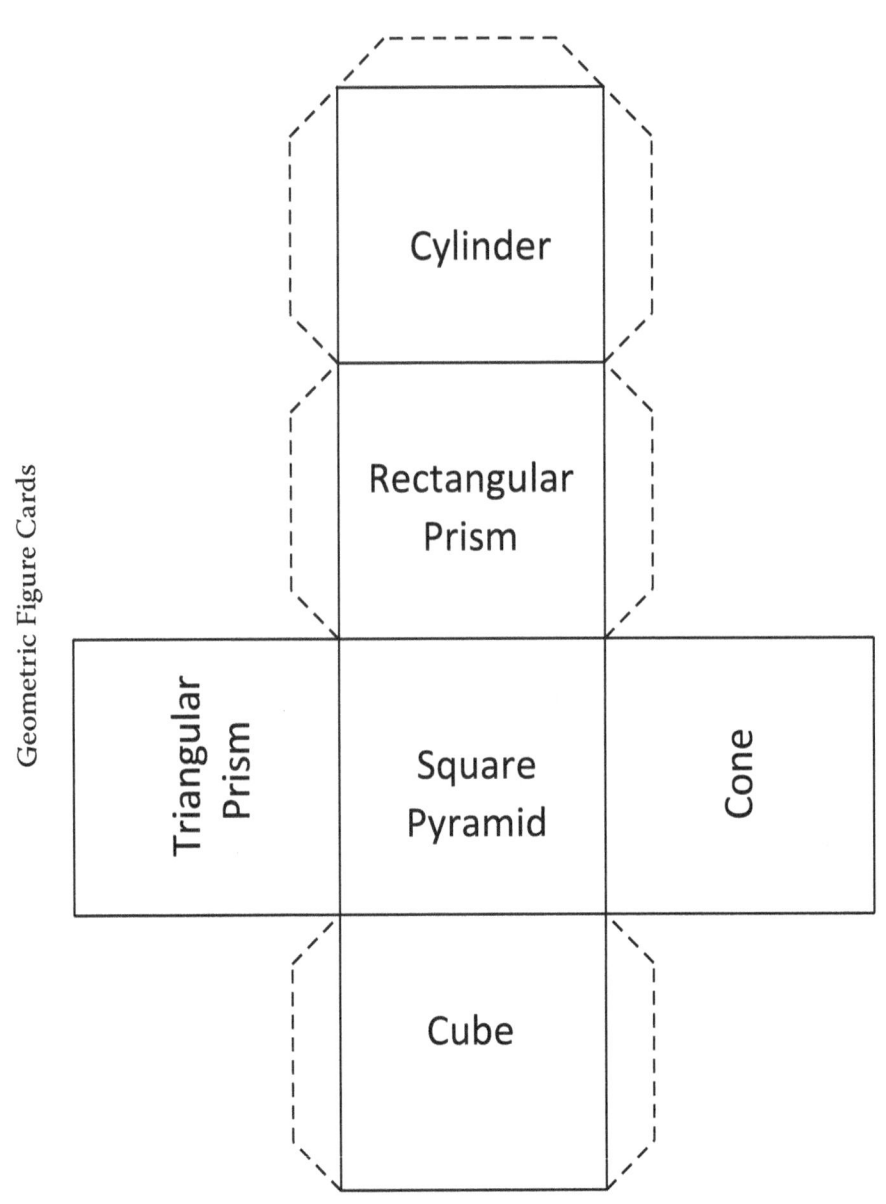

Geometric Figure Cards

Two-Dimensional

Compose	Compose	Compose
Two triangles and a square	A triangle, a square, and a rectangle	A trapezoid, a square, and a triangle
Compose	**Compose**	**Compose**
A pentagon, a rectangle, and a triangle	Three triangles and a hexagon	A pentagon, a square, and a triangle
Compose	**Compose**	**Compose**
Three distinct triangles	A square, a hexagon, and a triangle	A square and two rectangles

Compose	**Compose**	**Compose**
A triangle, a rectangle, and a semicircle.	A triangle and a circle	A rectangle and two circles
Compose	**Compose**	**Compose**
A square and four semicircles	A trapezoid and a semicircle	A circle and a square
Compose	**Compose**	**Compose**
A pentagon and a semicircle	A hexagon and two semicircles	A triangle and a semicircle

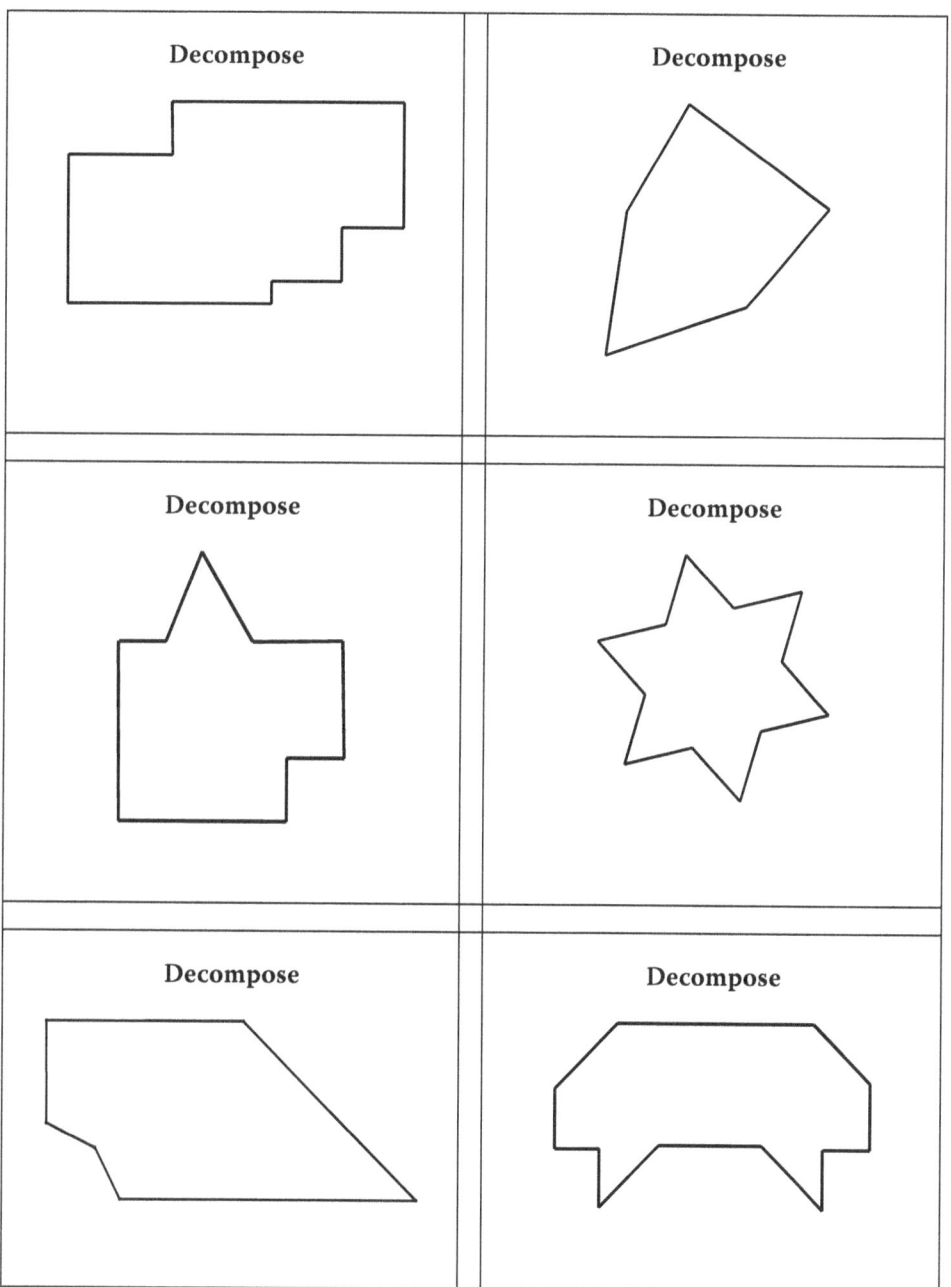

Decompose	Decompose
Decompose	Decompose
Decompose	Decompose

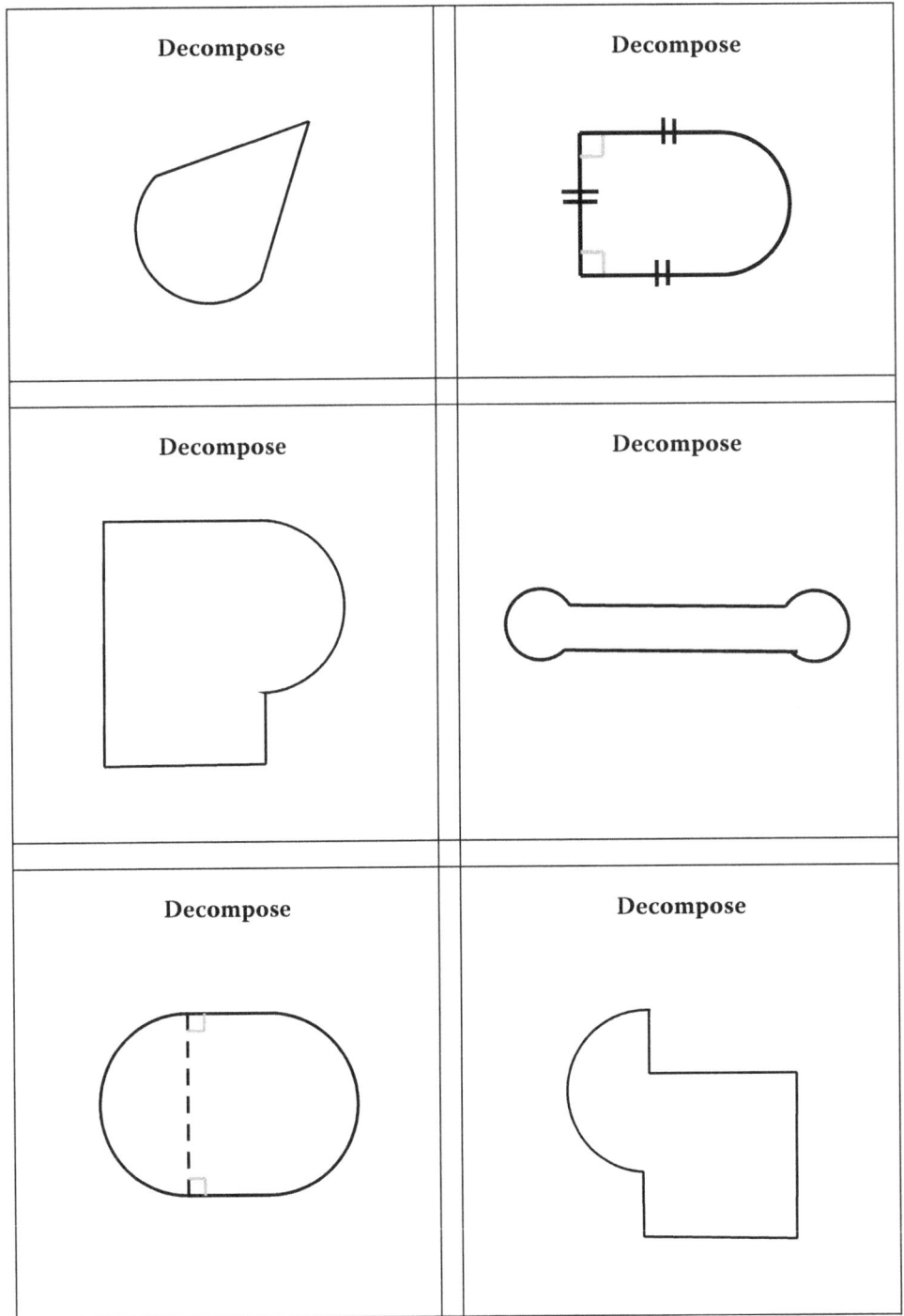

Geometric Figure Cards

Three-Dimensional

A cube and a pyramid	A rectangular prism and a triangular prism	Two rectangular prisms
Two cubes and a rectangular prism	A hexagonal prism and a tetrahedron	Two tetrahedrons and a cube
Three triangular prisms	A pentagonal prism and a triangular prism	Three rectangular prisms

A cylinder and a cone	A cylinder and a hemisphere	Two cones, a rectangular prism, and four cubes
A rectangular prism and two hemispheres	A sphere and a cone	A cylinder and two spheres
A cone and a cylinder	A cube, a cylinder, and a cone	Two cones

Geometric Figure Cards

Create a Figure

Draw an isosceles trapezoid.	Draw an obtuse right triangle.	Draw a quadrilateral with side lengths 9 cm, 5 cm, 4 cm, and 6 cm.
Draw a quadrilateral with the following angle measures: 100°, 90°, 100°, and 80°.	Draw a quadrilateral with two angles measuring 60°.	Draw a triangle with side lengths 1 cm, 1 cm, and 2 cm.
Draw a quadrilateral with no sides congruent and no sets of sides parallel.	Draw a quadrilateral with adjacent sides perpendicular.	Draw a pentagon with all sides congruent.

Draw a 3-D shape that has a circular cross section.	Draw a 3-D shape that has a square cross section.	Draw a 3-D shape that has all edges congruent and whose corners meet in obtuse angles.
Draw a 3-D shape whose cross section resembles a parallelogram.	Draw a 3-D shape whose cross section resembles a trapezoid.	Draw a 3-D shape whose cross section resembles a cube.
Draw a 3-D shape whose cross section could be either a rectangle or a triangle.	Draw a 3-D shape whose cross section is always a circle.	Draw a 3-D shape whose cross section could be either a circle or a triangle.

Geometric Figure Cards

2-D Scale Drawing

A textbook	A bulletin board, white board, etc., in your classroom	A pencil, pen, or marker
A bookcase	A door in the room	A basketball court floor
A cereal box	A wall in your classroom	A soda can

CHAPTER 10

RAFT

The RAFT (Role, Audience, Format, Topic) writing strategy allows students to utilize their creativity within a mathematical context. Students are given a situation, along with a structure, which sets the parameters for the piece. If using as a formative assessment, the topic can be kept vague, as the bold text indicates, to determine whether students have internalized the key understanding(s) of the topic. If the intention is to ensure specific mathematical points are included in the piece, clearly state that expectation within the topic guidelines as the additional text demonstrates.

Role	Audience	Format	Topic
Integer	Number line	Directions	**Find My Location:** Explain how to locate an integer on the number line.
Expression	Equation	Partnership agreement	**Our Collaboration:** Discuss how expressions and equations work together.
Fraction	Operation signs	Cartoon	**Drawing Conclusions:** You don't always increase our value!

DOI: 10.4324/9781003374589-12

Role	Audience	Format	Topic
Circumference of a circle	Area of a circle	Invitation to a family reunion	**How We Are Related:** Explain the relationship between the circumference and area of a circle.
Area	|Two-dimensional (2-D) figure	Missing persons ad	**Looking For:** Describe how to calculate the area of a figure.
System of equations	Jury	Instructions	**Finding Solutions:** Discuss how to determine if there is a solution and, if so, how many.
Function	Nonfunction	Breakup letter	**Irreconcilable Differences:** Explain the difference between a function and a nonfunction.
Parallel lines	Transversal	Last will and testament	**Division of Assets:** Explain the angles formed by cutting parallel lines with a transversal.
Zero	Proportional relationship	Social media post	**Relationship Status:** Discuss the role of zero in a proportional relationship.
Student choice	Student choice	Student choice	Student choice

For a problem-based experience, a RAFT can also be incorporated within a larger task as in the next example. An element of questioning has been added to reinforce its importance in problem-solving as well as provide a scaffold for entry. Once students have generated questions for the client, reveal that, for this problem, the client has stated they are unsure of the number of people attending so the students will need to determine which company is the best buy for all possible cases.

Role	What is the writer's role?	Event planner
Audience	Who will be reading the piece?	Client
Format	What is the best way to present the information?	Proposal
Topic	Who or what is the subject?	Best buy for printing tickets for event

What questions do you have for the client?

Three companies have submitted bids. Their pricing structures are outlined below:

Print Pro	Deluxe Printing	Printing Plaza
$15 setup fee plus $0.04 per ticket	250 tickets for $25	0–199 tickets $0.20/ticket 200–499 tickets $0.10/ticket 500 or more $0.05/ticket

PROPOSAL
Client:
Scope of Work:
Company Comparisons:
Cost Analysis:
Recommendation:

Additional Format Ideas

Speech	Journal Entry	Script	Song
Public Service Announcement	Trading Card	Story Board	Advice Column
Text Message	Petition	Letter (apology, persuasive, thank you, complaint)	Commercial
Advertisement	Recipe	Campaign Speech	Wanted Poster
Itinerary	Personal Ad	Nursery Rhyme/ Riddle	Infographic
Editorial	Eulogy	Current Event	Debate

Additional Samples

<u>Role:</u> Calculator
<u>Audience:</u> Math student
<u>Format:</u> Directions
<u>Topic:</u> Don't use me to multiply and divide by magnitudes of 10!

<u>Role:</u> Square
<u>Audience:</u> Cube
<u>Format:</u> Letter
<u>Topic:</u> You are nothing without me. (Include discussions of one-dimensional (1-D), 2-D, and three-dimensional (3-D) objects.)

<u>Role:</u> Negative integer
<u>Audience:</u> Positive integers
<u>Format:</u> Dating app input
<u>Topic:</u> Opposites attract!

<u>Role:</u> Rational numbers
<u>Audience:</u> Irrational numbers
<u>Format:</u> Rap
<u>Topic:</u> Must you go on forever?

CHAPTER 11

Question Quilt

The question quilt can be a strategy for differentiation that allows for student agency as they read and think about questions and statements relating to a topic of study. They then choose a few of the boxes to think about further. Students decide if they agree or disagree with the statement(s) and/or answer the question(s) and write justification supporting their responses. Questions and statements can be framed to accommodate a variety of levels of learners.

Another option is to give students the question quilt as you are beginning a topic and they can discuss the questions as they progress through the development of the topic.

Sample directions: Choose at least three questions or statements from the question quilt. Answer the question or decide if you agree or disagree with the statement. Justify your responses fully.

DOI: 10.4324/9781003374589-13

Question Quilt
Ratios and Proportional Reasoning

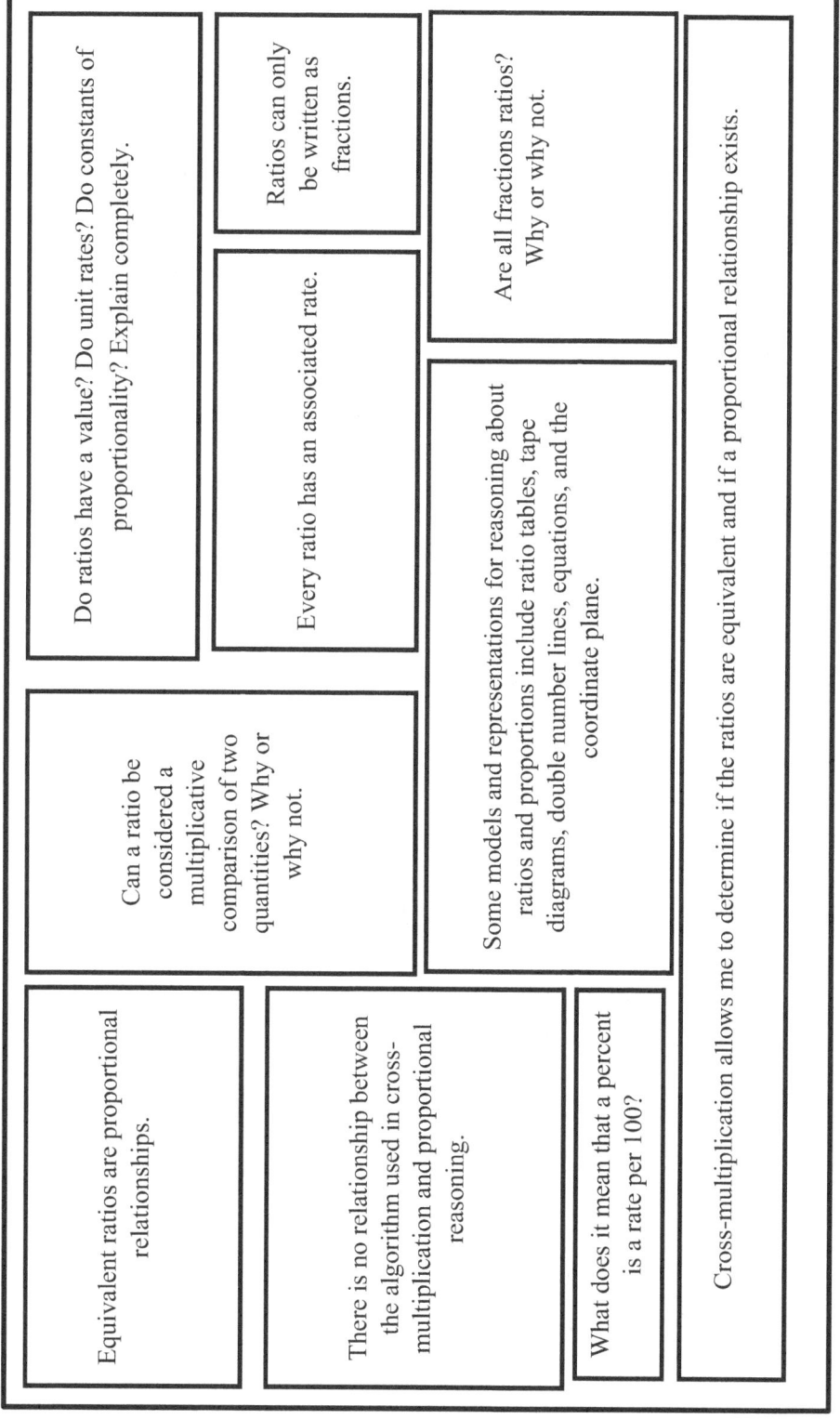

Do ratios have a value? Do unit rates? Do constants of proportionality? Explain completely.

Ratios can only be written as fractions.

Are all fractions ratios? Why or why not.

Every ratio has an associated rate.

Some models and representations for reasoning about ratios and proportions include ratio tables, tape diagrams, double number lines, equations, and the coordinate plane.

Can a ratio be considered a multiplicative comparison of two quantities? Why or why not.

Equivalent ratios are proportional relationships.

There is no relationship between the algorithm used in cross-multiplication and proportional reasoning.

What does it mean that a percent is a rate per 100?

Cross-multiplication allows me to determine if the ratios are equivalent and if a proportional relationship exists.

Question Quilt
Ratios and Proportional Reasoning Answer Key

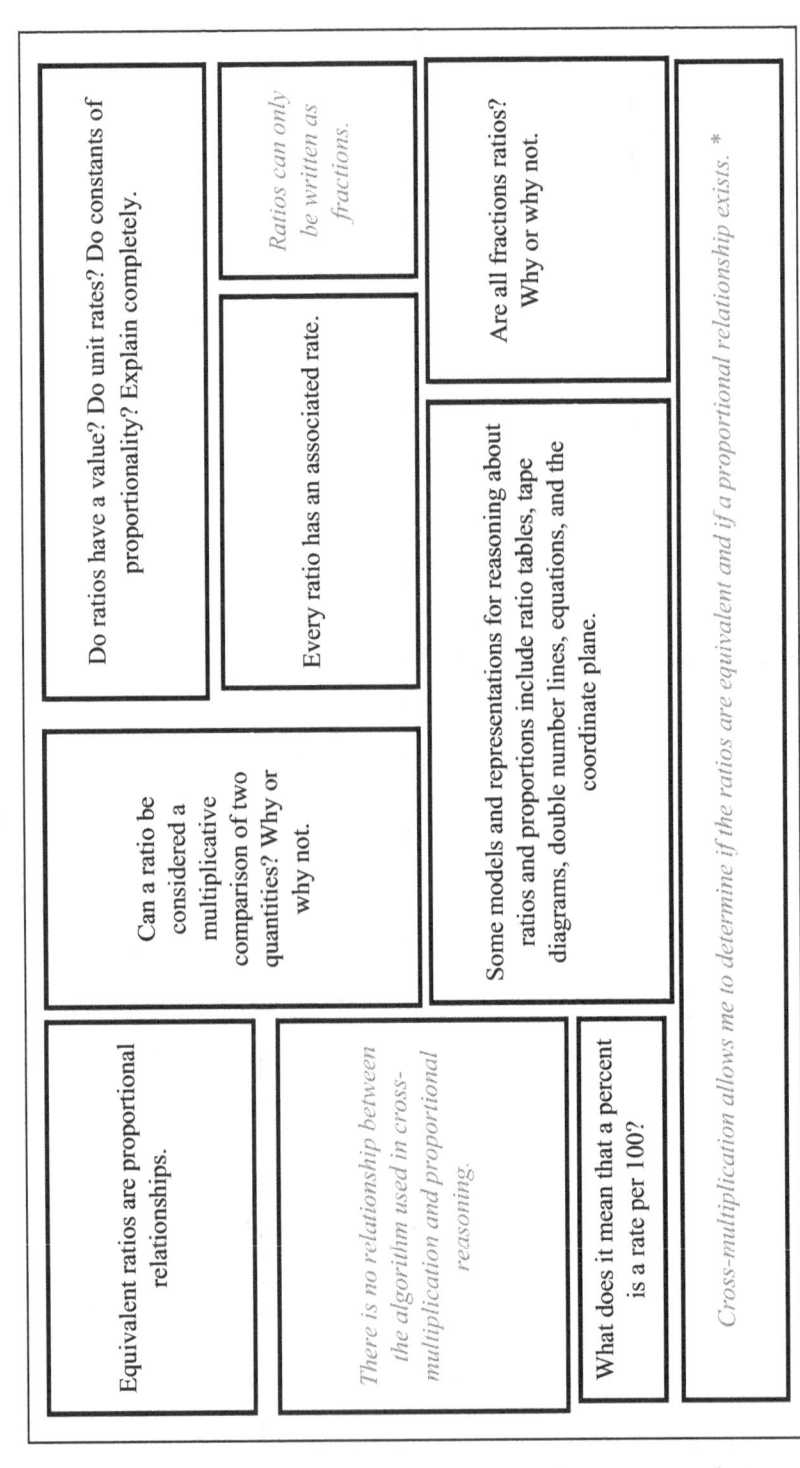

Do ratios have a value? Do unit rates? Do constants of proportionality? Explain completely.

Ratios can only be written as fractions.

Are all fractions ratios? Why or why not.

Every ratio has an associated rate.

Can a ratio be considered a multiplicative comparison of two quantities? Why or why not.

Some models and representations for reasoning about ratios and proportions include ratio tables, tape diagrams, double number lines, equations, and the coordinate plane.

Equivalent ratios are proportional relationships.

There is no relationship between the algorithm used in cross-multiplication and proportional reasoning.

What does it mean that a percent is a rate per 100?

*Cross-multiplication allows me to determine if the ratios are equivalent and if a proportional relationship exists. ***

*The cross-multiplication algorithm can only be used to determine whether fractions are equivalent. It CANNOT be used to determine whether the fractions are proportional. This can only be determined by the context of the situation and understanding of the relationships present. NOTE: Therefore, we say the fractions are equivalent, not the ratios. There must be a context given to have a ratio.

Question Quilt

Expressions, Equations, and Functions

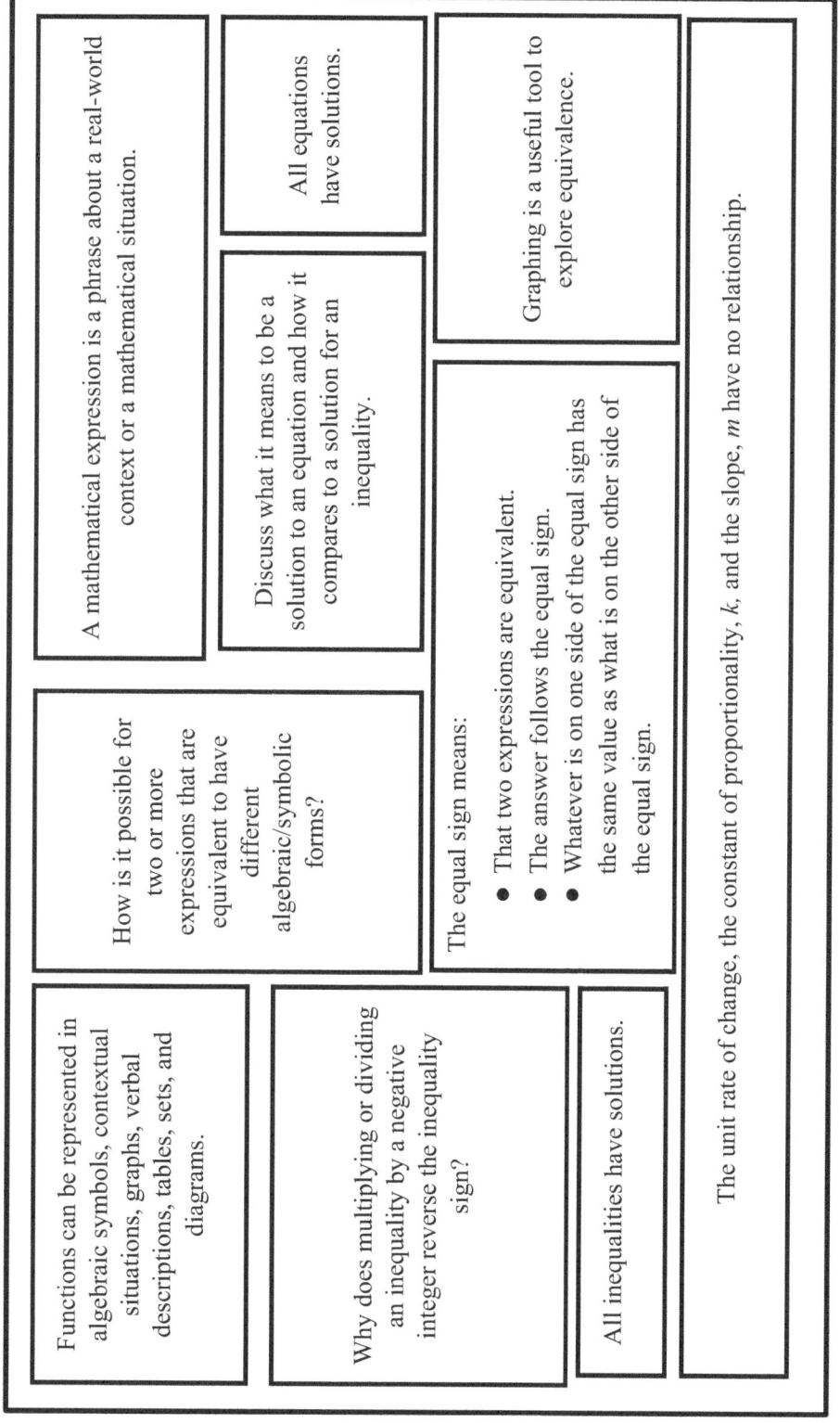

A mathematical expression is a phrase about a real-world context or a mathematical situation.

All equations have solutions.

Discuss what it means to be a solution to an equation and how it compares to a solution for an inequality.

Graphing is a useful tool to explore equivalence.

How is it possible for two or more expressions that are equivalent to have different algebraic/symbolic forms?

The equal sign means:

- That two expressions are equivalent.
- The answer follows the equal sign.
- Whatever is on one side of the equal sign has the same value as what is on the other side of the equal sign.

Functions can be represented in algebraic symbols, contextual situations, graphs, verbal descriptions, tables, sets, and diagrams.

Why does multiplying or dividing an inequality by a negative integer reverse the inequality sign?

All inequalities have solutions.

The unit rate of change, the constant of proportionality, k, and the slope, m have no relationship.

Question Quilt
Expressions, Equations, and Functions Answer Key

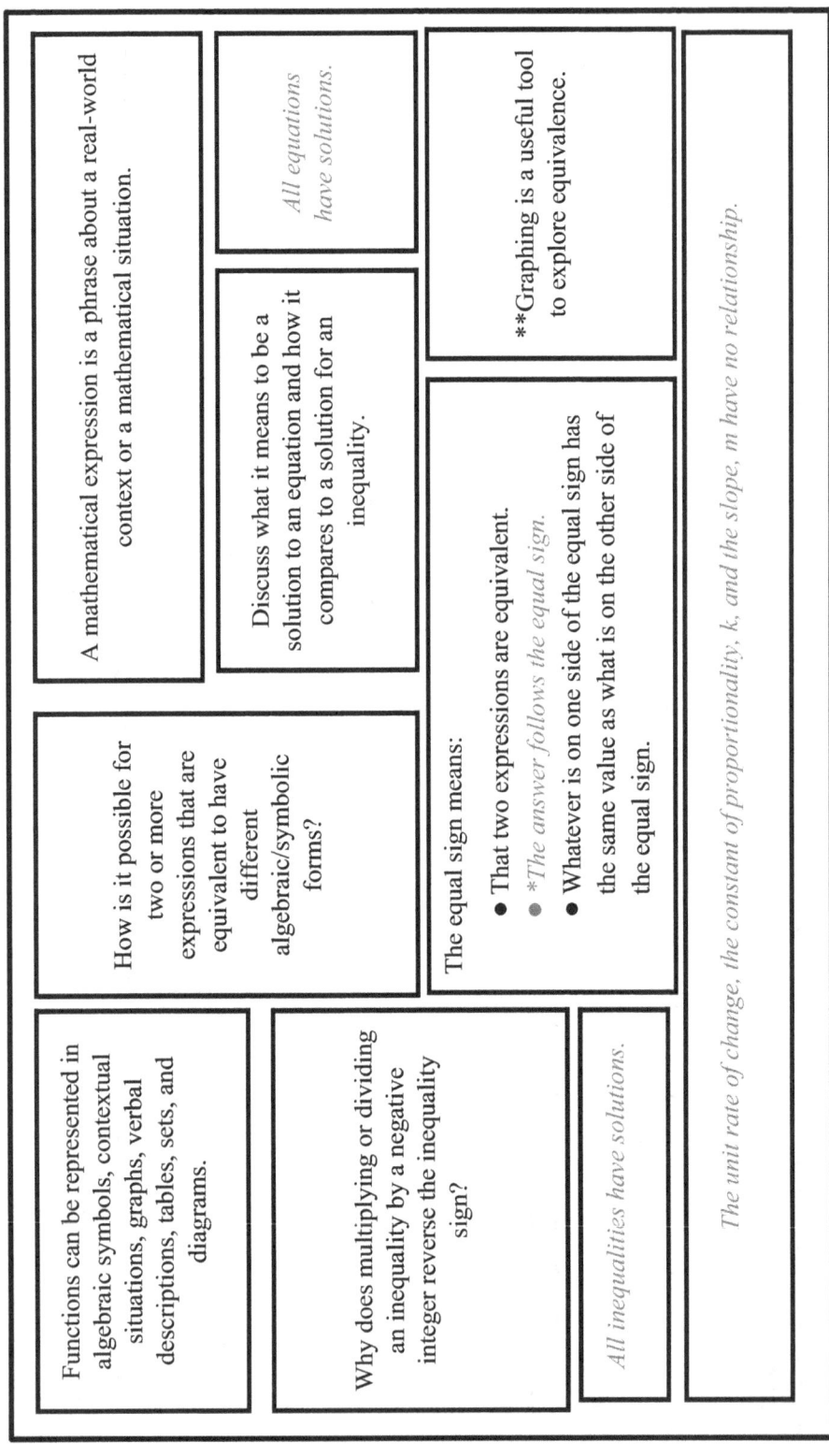

A mathematical expression is a phrase about a real-world context or a mathematical situation.

All equations have solutions.

Discuss what it means to be a solution to an equation and how it compares to a solution for an inequality.

**Graphing is a useful tool to explore equivalence.

How is it possible for two or more expressions that are equivalent to have different algebraic/symbolic forms?

The equal sign means:
● That two expressions are equivalent.
● *The answer follows the equal sign.*
● Whatever is on one side of the equal sign has the same value as what is on the other side of the equal sign.

Functions can be represented in algebraic symbols, contextual situations, graphs, verbal descriptions, tables, sets, and diagrams.

Why does multiplying or dividing an inequality by a negative integer reverse the inequality sign?

All inequalities have solutions.

The unit rate of change, the constant of proportionality, k, and the slope, m have no relationship.

* The equal sign does not always indicate the answer follows.
**Graphing can be used to explore equivalence with respect to expressions, equations, and systems of equations

CHAPTER 12

Always, Sometimes, and Never

Using always, sometimes, and never questions provides the opportunity for students to investigate a statement and determine whether it is true all the time, some of the time, or never. Typically, always, sometimes, and never have been associated with geometry, but as you can see from the following pages, these are as easily adapted to other topics. They are rich tasks because they engage students in a higher level of reasoning and communication than often typical questions might. This activity allows teachers a look into the thinking of the students.

Students may not simply write A, S, or N. They must validate and support their choice. If they determine the statement is sometimes, students should be required to offer a case for when the statement does hold (an example) as well as when it does not (a counterexample). Many true or false questions can easily be adapted for always, sometimes, and never statements.

DOI: 10.4324/9781003374589-14

Sample Statements

Topic 1: Number and Quantity

1. When you multiply two numbers, you get a larger number.
2. When you divide two numbers, you get a smaller number.
3. A rational number added with an irrational number is irrational.
4. Multiplying even numbers of integers results in a positive product.
5. Dividing by zero gives a quotient that is a real number.
6. Fractions are greater than whole numbers.
7. The length 7.045 cm is longer than the length 7.45 cm.
8. Inverses operate to give identities.
9. Every real number has a decimal expansion.
10. The sum of four consecutive whole numbers is divisible by four.

Topic 2: Algebraic Reasoning

Expressions

1. All equations have solutions.
2. When you divide an inequality by a negative integer, the inequality sign changes direction.
3. $-x^2$ is a positive integer.
4. $2(x-3) = 2x - 6$
5. $5x \geq 8 + x$
6. $x^2 = x$
7. $x^2 > x^3$
8. $\dfrac{2x+4}{2} = x + 2$
9. Division of integers is commutative.
10. A system of lines that have the same slope will intersect at one point.

Equations

1. $6x - 6 = 6(x+1)$
2. $5x = 5$
3. $4 - x = x - 4$
4. $\dfrac{4x+8}{4} = x + 2$

5. Some equations have no solution.
6. Solutions to equations can be checked using substitution.
7. Parallel lines have equal slopes.
8. In a linear relationship, if you double a domain value, the corresponding range value is also doubled.
9. When solving an equation involving the distributive property, the order in which you deal with the coefficient on the parentheses and the constant inside the parentheses does not matter. For example, $5(x-2)=10$; it does not matter whether I first divide by 5 or add 2.
10. The graph of $y = 3$ intersects the x-axis.

Inequalities

1. $5x > 5 + x$
2. $n + 4 > 8$
3. $|2x+3| < 0$
4. $x \leq 7$ is equivalent to $x = 7$.
5. Inequalities have more than one solution.
6. When you graph an inequality, there is a shaded representation for the solution(s).
7. The conditionals, "and" and "or," represent the same constraint when working with inequalities.
8. "At least" and "at most" imply that the equal sign must be used with the inequality sign.
9. "Less than" signifies the left side of the number line.
10. The properties of equality apply in the same way when working with inequalities.

Functions

1. Functions are relations.
2. Relations are functions.
3. Functions are vertical lines.
4. Functions can be evaluated.
5. Input, domain, and x-value are equivalent concepts.
6. Functions are linear.
7. Tables and graphs show different attributes of a function from the algebraic form.
8. A context is not needed to fully make sense of a solution to a function.
9. The rate of change does not reveal the relationship between the two attributes defined by the function rule.
10. Functions are horizontal lines.

Systems

1. A system of equations has one solution.
2. A system of equations consists of two equations.
3. You can determine if a system of equations has a solution through graphing.
4. Two lines that are not parallel can have multiple solutions.
5. The solution to a linear system is where the lines intersect.
6. Systems of equations can only be solved using substitution and/or elimination.
7. If the final step when solving a system shows "$-7 = -7$," it implies there is no solution.
8. If a solution to a linear system is an ordered pair, the lines have different slopes.
9. The points of intersection that make up the solution to a linear system satisfy both equations simultaneously.
10. The point $(1,-5)$ is a solution to the system $3x - 2y = 5$ and $2x - y = 7$.

Topic 3: Geometric Reasoning/Measurement and Units

Quadrilaterals

1. If two rectangles have the same perimeter, they have the same area.
2. If you join the midpoints of the sides of a quadrilateral, the resulting figure is a parallelogram with the area one-half the area of the original quadrilateral.
3. When you cut a piece off a quadrilateral, you reduce its perimeter and area.
4. Rectangles are squares.
5. Trapezoids are parallelograms.
6. Diagonals of a parallelogram are congruent.
7. Diagonals of a rhombus are perpendicular.
8. A quadrilateral is a regular polygon.
9. A parallelogram has exactly three right angles.
10. A square is a rhombus.

Circles

1. Circles are polygons.
2. Radii in a circle are congruent.
3. The circumference of a circle is a whole number.
4. All circles are similar.

5. The relationship between the area of a circle and the diameter of a circle involves the irrational number, π.
6. If a shape has an area of 16π, the shape is a circle.
7. Three points determine a circle.
8. Chords of a circle are longer than the radius of a circle.
9. A chord of a circle is the diameter of the circle.
10. The center of a circle is a point on the circle.

Triangles

1. A triangle has exactly two lines of symmetry.
2. A scalene triangle can be isosceles.
3. An equilateral triangle can be equiangular.
4. Regular polygons have an equal number of sides.
5. Triangles can have more than one obtuse angle.
6. Right triangles are isosceles.
7. Acute triangles are right triangles.
8. Equilateral triangles are similar.
9. Scalene triangles have two acute angles.
10. Exterior angles of right triangles are acute.

Transformations

1. A composition of isometries is an isometry.
2. Composition of transformations is commutative.
3. A rotation is an isometry.
4. A rotation followed by another rotation can be written as a single rotation.
5. A translation followed by a reflection has a point that does not change position.
6. If an angle is dilated with the center of dilation at the vertex, the angle measure may change.
7. Two figures are congruent when there is a sequence of transformations that match one figure with the other.
8. If the length of a right rectangular prism is doubled, the surface area is doubled.
9. The translation image of any shape can be sketched given a point-to-point translation.
10. A scaled image of any shape can be sketched given a center of rotation.

Congruence and Similarity

1. Similar triangles are also congruent.
2. Congruent triangles are similar.
3. Similar figures can be different sizes.
4. Similar figures can be different shapes.
5. Corresponding side lengths of similar figures are proportional.
6. Corresponding angles of similar figures are congruent.
7. Equilateral triangles are congruent.
8. The distances between pairs of corresponding points in congruent figures are not equal.
9. The ratio of corresponding sides of similar polygons is constant.
10. Figures that are congruent have the same area.
11. If you draw two shapes, the shape with the greater area will also have the longer perimeter.
12. If a square and a circle have the same perimeter/circumference, the circle has the smaller area.
13. Figures that have equal areas are congruent.

Topic 4: Data Analysis, Probability, and Statistics

Data Analysis and Statistics:

1. The median is the center based upon the frequency of the data set.
2. The mean is included in the data set.
3. The largest element in the data set is the mode of the data set.
4. The mean can be a negative number.
5. An outlier is the data point to the far left on a line plot.
6. Every data set has a mode.
7. Statistical questions have more than one answer.
8. A distribution that is skewed left has a left tail that is longer than the right tail.
9. Histograms are like bar graphs and have spaces between each vertical bar.
10. Measures of central tendency can be determined for a bivariate data set.

Probability

1. Scoring a sum of "3" with two number cubes is twice as likely as scoring a sum of "2."
2. The experimental probability of getting exactly two heads tossing four two-sided coins is one-half.

3. The probability of an event happening can be greater than 1.
4. The theoretical probability of an event happening is the same as the experimental probability of the same event happening.
5. The conjunction "and," when used in probability, means to add.
6. The probability of a compound event is the percentage of outcomes in the sample space for which the compound event occurs.
7. Tree diagrams and tables are the only way to represent sample spaces for compound events.
8. The probability of it snowing tomorrow is .5 because it will either snow or not snow.
9. A probability represented by a percentage is different from a probability represented by the equivalent fraction.
10. The basic definition for probability can be represented by the ratio $\dfrac{n(E)}{n(S)}$, where n is the number of E (events) divided by the number in the S (sample space).

Sample Statements with Answers

Topic 1: Number and Quantity

11. When you multiply two numbers, you get a larger number. **(S)**
12. When you divide two numbers, you get a smaller number. **(S)**
13. A rational number added with an irrational number is irrational. **(A)**
14. Multiplying even numbers of integers results in a positive product. **(S)**
15. Dividing by zero gives a quotient that is a real number. **(N)**
16. Fractions are greater than whole numbers. **(S)**
17. The length 7.045 cm is longer than the length 7.45 cm. **(N)**
18. Inverses operate to give identities. **(A)**
19. Every real number has a decimal expansion. **(A)**
20. The sum of four consecutive whole numbers is divisible by four. **(N)**

Topic 2: Algebraic Reasoning

Expressions

11. All equations have solutions. **(S)**
12. When you divide an inequality by a negative integer, the inequality sign changes direction. **(A)**
13. $-x^2$ is a positive integer. **(N)**

14. $2(x-3)=2x-6$ **(A)**
15. $5x \geq 8+x$ **(S)**
16. $x^2 = x$ **(S)**
17. $x^2 > x^3$ **(S)**
18. $\dfrac{2x+4}{2} = x+2$ **(A)**
19. Division of integers is commutative. **(N)**
20. A system of lines that have the same slope will intersect at one point. **(N)**

Equations

11. $6x-6=6(x+1)$ **(N)**
12. $5x=5$ **(S)**
13. $4-x=x-4$ **(S)**
14. $\dfrac{4x+8}{4} = x+2$ **(A)**
15. Some equations have no solution. **(A)**
16. Solutions to equations can be checked using substitution. **(S)**
17. Parallel lines have equal slopes. **(A)**
18. In a linear relationship, if you double a domain value, the corresponding range value is also doubled. **(A)**
19. When solving an equation involving the distributive property, the order in which you deal with the coefficient on the parentheses and the constant inside the parentheses does not matter. For example, $5(x-2)=10$; it does not matter whether I first divide by 5 or add 2. **(N)**
20. The graph of $y=3$ intersects the x-axis. **(N)**

Inequalities

11. $5x > 5 + x$ **(S)**
12. $n+4 > 8$ **(S)**
13. $|2x+3| < 0$ **(N)**
14. $x \leq 7$ is equivalent to $x=7$ **(N)**
15. Inequalities have more than one solution. **(S)**
16. When you graph an inequality, there is a shaded representation for the solution(s). **(A)**
17. The conditionals, "and" and "or," represent the same constraint when working with inequalities. **(N)**
18. "At least" and "at most" imply that the equal sign must be used with the inequality sign. **(A)**

19. "Less than" signifies the left side of the number line. **(A)**
20. The properties of equality apply in the same way when working with inequalities. **(S)**

Functions

11. Functions are relations. **(A)**
12. Relations are functions. **(S)**
13. Functions are vertical lines. **(N)**
14. Functions can be evaluated. **(A)**
15. Input, domain, and x-value are equivalent concepts. **(A)**
16. Functions are linear. **(S)**
17. Tables and graphs show different attributes of a function from the algebraic form. **(A)**
18. A context is not needed to fully make sense of a solution to a function. **(N)**
19. The rate of change does not reveal the relationship between the two attributes defined by the function rule. **(N)**
20. Functions are horizontal lines. **(S)**

Systems

11. A system of equations has one solution. **(S)**
12. A system of equations consists of two equations. **(S)**
13. You can determine if a system of equations has a solution through graphing. **(A)**
14. Two lines that are not parallel can have multiple solutions. **(N)**
15. The solution to a linear system is where the lines intersect. **(A)**
16. Systems of equations can only be solved using substitution and/or elimination. **(N)**
17. If the final step when solving a system shows "$-7 = -7$," it implies there is no solution. **(N)**
18. If a solution to a linear system is an ordered pair, the lines have different slopes. **(A)**
19. The points of intersection that make up the solution to a linear system satisfy both equations simultaneously. **(A)**
20. The point $(1, -5)$ is a solution to the system $3x - 2y = 5$ and $2x - y = 7$. **(N)**

Topic 3: Geometric Reasoning/Measurement and Units

Quadrilaterals

11.   If two rectangles have the same perimeter, they have the same area. **(S)**
12.   If you join the midpoints of the sides of a quadrilateral, the resulting figure is a parallelogram with the area one-half the area of the original quadrilateral. **(A)**
13.   When you cut a piece off a quadrilateral, you reduce its perimeter and area. **(S)**
14.   Rectangles are squares. **(S)**
15.   Trapezoids are parallelograms. **(N)** (Disclaimer: This assumes the family of quads is sorted by zero sets of sides parallel, one set of sides parallel, and two sets of sides parallel.)
16.   Diagonals of a parallelogram are congruent. **(A)**
17.   Diagonals of a rhombus are perpendicular. **(A)**
18.   A quadrilateral is a regular polygon. **(S)**
19.   A parallelogram has exactly three right angles. **(N)**
20.   A square is a rhombus. **(A)**

Circles

11.   Circles are polygons. **(N)**
12.   Radii in a circle are congruent. **(A)**
13.   The circumference of a circle is a whole number. **(N)**
14.   All circles are similar. **(A)**
15.   The relationship between the area of a circle and the diameter of a circle involves the irrational number, π. **(A)**
16.   If a shape has an area of 16π, the shape is a circle. **(S)**
17.   Three points determine a circle. **(S)**
18.   Chords of a circle are longer than the radius of a circle. **(S)**
19.   A chord of a circle is the diameter of the circle. **(S)**
20.   The center of a circle is a point on the circle. **(N)**

Triangles

11.   A triangle has exactly two lines of symmetry. **(N)**
12.   A scalene triangle can be isosceles. **(S)**
13.   An equilateral triangle can be equiangular. **(A)**
14.   Regular polygons have an equal number of sides. **(S)**
15.   Triangles can have more than one obtuse angle. **(N)**
16.   Right triangles are isosceles. **(S)**

17. Acute triangles are right triangles. (**S**)
18. Equilateral triangles are similar. (**A**)
19. Scalene triangles have two acute angles. (**A**)
20. Exterior angles of right triangles are acute. (**N**)

Transformations

11. A composition of isometries is an isometry. (**A**)
12. Composition of transformations is commutative. (**S**)
13. A rotation is an isometry. (**A**)
14. A rotation followed by another rotation can be written as a single rotation. (**A**)
15. A translation followed by a reflection has a point that does not change position. (**S**)
16. If an angle is dilated with the center of dilation at the vertex, the angle measure may change. (**N**)
17. Two figures are congruent when there is a sequence of transformations that match one figure with the other. (**A**)
18. If the length of a right rectangular prism is doubled, the surface area is doubled. (**S**)
19. The translation image of any shape can be sketched given a point-to-point translation. (**A**)
20. A scaled image of any shape can be sketched given a center of rotation. (**N**)

Congruence and Similarity

14. Similar triangles are also congruent. (**S**)
15. Congruent triangles are similar. (**A**)
16. Similar figures can be different sizes. (**S**)
17. Similar figures can be different shapes. (**N**)
18. Corresponding side lengths of similar figures are proportional. (**A**)
19. Corresponding angles of similar figures are congruent. (**A**)
20. Equilateral triangles are congruent. (**S**)
21. The distance between pairs of corresponding points in congruent figures are not equal. (**N**)
22. The ratio of corresponding sides on similar polygons is constant. (**A**)
23. Figures that are congruent have the same area. (**A**)
24. If you draw two shapes, the shape with the greater area will also have the longer perimeter. (**S**)
25. If a square and a circle have the same perimeter/circumference, the circle has the smaller area. (**S**)
26. Figures that have equal areas are congruent. (**S**)

Topic 4: Data Analysis, Probability, and Statistics

Data Analysis and Statistics

11. The median is the center based upon the frequency of the data set. **(N)**
12. The mean is included in the data set. **(S)**
13. The largest element in the data set is the mode of the data set. **(S)**
14. The mean can be a negative number. **(S)**
15. An outlier is the data point to the far left on a line plot. **(S)**
16. Every data set has a mode. **(S)**
17. Statistical questions have more than one answer. **(A)**
18. A distribution that is skewed left has a left tail that is longer than the right tail. **(A)**
19. Histograms are like bar graphs and have spaces between each vertical bar. **(N)**
20. Measures of central tendency can be determined for a bivariate data set. **(N)**

Probability

11. Scoring a sum of "3" with two number cubes is twice as likely as scoring a sum of "2." **(A)**
12. The experimental probability of getting exactly two heads tossing four two-sided coins is one-half. **(S)**
13. The probability of an event happening can be greater than 1. **(N)**
14. The theoretical probability of an event happening is the same as the experimental probability of the same event happening. **(S)**
15. The conjunction "and," when used in probability, means to add. **(N)**
16. The probability of a compound event is the percentage of outcomes in the sample space for which the compound event occurs. **(A)**
17. Tree diagrams and tables are the only way to represent sample spaces for compound events. **(N)**
18. The probability of it snowing tomorrow is .5 because it will either snow or not snow. **(S)**
19. A probability represented by a percentage is different from a probability represented by the equivalent fraction. **(N)**
20. The basic definition for probability can be represented by the ratio $\frac{n(E)}{n(S)}$, where n is the number of E (events) divided by the number in the S (sample space). **(A)**

PART THREE

Planning and Implementation

CHAPTER 13

Crosswalks

The following crosswalk is included to support instructional planning. Resources can be quickly identified based upon the mathematical topic as well as the type of writing and/or the strategy example given. The crosswalk identifies the mathematical topics that are included in the given examples referenced in each of the 11 writing strategies shared in Chapter 12.

DOI: 10.4324/9781003374589-16

Crosswalk of Topics and Writing Strategies

Writing Strategy \ Topics	Number and Quantity	Algebraic Reasoning	Geometric Reasoning/ Measurement and Units	Data Analysis, Probability, and Statistics	Universal
Always, Sometimes, and Never	X	X	X	X	
Question Quilts	X	X			
R.A.F.T.	X	X	X		
Cubing/ Think Dots		X	X		
Poems	X		X		X
Journal Prompts	X	X	X		X
Writing About	X	X	X	X	
Topical Questions	X	X	X	X	
The Answer Is...	X	X	X	X	
Compare/ Contrast	X	X	X	X	X
Visual Prompts	X	X	X		

CHAPTER 14

Bringing It All Together

This chapter provides a sample anchor task that demonstrates how several of these writing strategies can be authentically integrated into classroom instruction. A lesson plan and facilitation notes are provided.

Selling the Farm

Overview: This task was intentionally chosen because it not only models several opportunities for writing but also demonstrates the efficiency of planning for addressing multiple content and process standards. Planning for simultaneous outcomes allows time for students to dive deep into a single task rather than completing multiple tasks over the same timeframe. It allows students to make connections within the content rather than viewing the concepts as discrete and unrelated. This task is so versatile that it can be modified in unlimited ways to meet the needs of both teachers and students. The outline below is just one suggestion for how it can be facilitated.

DOI: 10.4324/9781003374589-17

Writing Opportunities:

- ❏ Think-Write-Pair-Share (p. 5)
- ❏ Topical questions: "We're Stuck"/"We're Done" (p. 13)
- ❏ Problem-solving process (p. 7)
- ❏ Reflection (p. 13)

Content Connections: (This task can be used across grades 6–8 so possible topics may include but are not limited to)

- ❏ Fractions
- ❏ Geometry –
    - ❏ Decomposing/Composing,
    - ❏ Scale (measuring of tangram pieces and determining scale),
    - ❏ Pythagorean Theorem (extension question related to dimensions of pieces)
- ❏ Percentages (based on fractional representation)
- ❏ Decimals
- ❏ Real-world context (connections to other disciplines – land value versus cost based on area)
- ❏ Unit analysis (acre)
- ❏ Problem-solving (standards for mathematical practice)

Routines

The five *practices for orchestrating productive mathematics discussions*

- ❏ Anticipating students' solutions to a mathematics task
- ❏ Monitoring students' in-class, "real-time" work on the task
- ❏ Selecting approaches and students to share them
- ❏ Sequencing students' presentations purposefully
- ❏ Connecting students' approaches and the underlying mathematics

Materials Needed

- ❏ Tangrams (see 'Preparation for Implementation: PLC Work)
- ❏ Visual of tangram square
- ❏ Dry erase markers
- ❏ Which One Doesn't Belong (WODB) Prompt
- ❏ Task handout (without embedded problem-solving process)
- ❏ Task handout (with embedded problem-solving process)
- ❏ "We're Stuck" and "We're Done" questions

- [ ] Student sample attempt #1
- [ ] Student sample attempt #2
- [ ] QUAD reflection

Preparation

For interactive tangrams (able to be written on and large enough to manipulate), use the blackline master to make tangrams out of card stock. Laminate and cut into pieces.

Complete the task as if you are a student. *Anticipate* student strategies, solutions, misconceptions, etc. This process informed the development of the "We're Stuck" and "We're Done" questions used to support students. See **Preparation for Implementation: PLC Work** for a more detailed explanation.

Warm Up: 10 Minutes

Display the image below and ask students to determine WODB and record their reasoning. Students can do this on an index card or whiteboard. The purpose is to ensure they commit to a choice and provide rationale. Designate four corners of the room and identify each space with the letters A, B, C, and D. Ask students to go to the corner of the room that corresponds to their choice. Have students discuss their reasoning and to select a student(s) to share these reasons with the rest of the class. Once all groups have shared, if there is a choice that was not selected, engage the class in determining what reasons could be used to make the case for that representation. Doing so reinforces to students that all choices are valid with reasons they do not belong. It also challenges students to look beyond the more obvious selections.

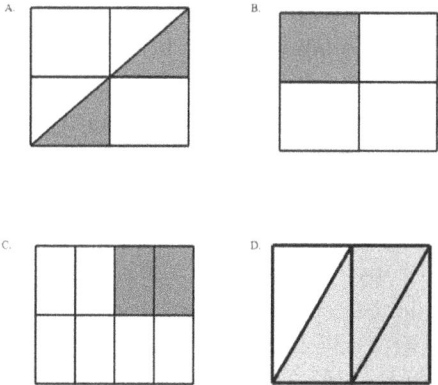

WODB

Some possible student responses may include, but are not limited to, the following:

A- It is the only one partitioned into six total parts.

B- It is the only one with a singular section shaded.

C- It is the only one that divided into eight equal parts.

D- It is the only one where the ¼ represents the unshaded versus the shaded portion.

Activity: 25 Minutes

Introduce the task by displaying it on the screen and reading aloud.

Distribute the handout and ask students to begin by working independently. Reinforce the expectations by displaying and stating the following:

Think about the problem

Write by doing one of the following:

❏ Write the question that is being asked and/or create an answer statement.

❏ Write all facts you know about the problem.

❏ Write what additional information is needed or questions you have about the problem.

Providing these questions allows access for all students and eliminates "no response" as an option. NOTE: If students are using the problem-solving process graphic organizer, there is a copy of the task handout that embeds the organizer for support in the workspace section.

Have students partner to share strategies and questions. This should be done intentionally using a collaborative structure and strategy to ensure equity in each pair with both students having a voice.

Students should record any additional ideas from the discussion on the handout.

Differentiation Strategy: As students are working, allow them to access to the scaffolded questions. This can be done by having students get them from a central location (hard copy) or access on a device (electronic copy). These can also be placed on student desks while monitoring and identifying where students may need support. For example, if students choose to use the tangrams and are having difficulty making the square, ask the first question provided on the "We're Stuck" sheet, "How does making the tangram square help in working this problem?" Once they can articulate why this is necessary, provide the blackline master of the square to allow students to focus on the problem rather than putting the pieces together. As students complete the task, monitor to make sure they are utilizing the "We're Done" questions.

Monitor student pairs and identify various strategies, misconceptions, "ah-ha" moments, etc., to highlight in the whole class discussion at the conclusion of the activity. *Select* (and inform) students prior to the discussion that they will be asked to share their thinking. Be specific in telling them what you would like them to share when called upon. This will keep the discussion focused on the main ideas.

Activity Synthesis: 20 Minutes

Conduct a classroom discussion by *sequencing* the student groups who were selected to share. This should be done in a manner that allows *connections* to be made between approaches and ideas to reinforce the learning goals set forth in the activity.

Student Reflection: 5 Minutes

Have students complete the QUAD Reflection independently.

- ❏ **Question:** What questions do you still have about the problem?
- ❏ **Understanding:** What do you now understand after working with this problem?
- ❏ **Activate:** How did this problem activate you as a learning resource for a peer or a peer for you?
- ❏ **Discourse:** What mathematical discourse was prompted by working this problem?

Preparation for Implementation: Professional Learning Community (PLC) Work

The best way to prepare for implementation of the task is for teachers to meet to unpack and discuss. Some ideas are outlined here.

1. Have all teachers complete the task as if they were students using the handout provided.

2. Discuss strategies and solutions from the group as well as other potential approaches. In addition, answer the questions below.

   a. Where will students demonstrate success with the task?
   b. Where will students struggle with the task? (OMG's- SREB abbreviation)

      ❏ Obstacles – Students have a lack of understanding of which strategies or procedures to apply and how those strategies work.
      ❏ Misconceptions – Students are unaware that the knowledge they have is incorrect.
      ❏ Gaps in Learning – Students lack prerequisite knowledge. (SREB, 2018)

3. Based on possible student struggles with the task, review the "stuck" questions that have been generated for support. Edit or add questions that may be needed. Keep in mind to limit the total number to between 3 and 5 so students don't get overwhelmed.

      ❏ How does making the tangram square help in working this problem?
      ❏ What is the relationship between each of the pieces of the tangram?
      ❏ What do you need to know to determine the value of each piece of land?
      ❏ What is the relationship between each of the pieces of the tangram and the whole tract of land? How will this help determine the value of each?

4. Complete the same process with the "done" questions.

      ❏ How can three different geometric shapes of land have the same value?
      ❏ Why did the values of each piece of land calculate to be whole numbers?
      ❏ How would the values change if the total price were not a multiple of the denominators of the fractional pieces?
      ❏ If the area of the tract of land measures 100 acres, what would be the area of each piece of land? What would be the dimensions of each piece? (IMPORTANT – This question is intentionally designed to cause students to think more deeply and is routed in an understanding of land measure. An acre is a unit of area and why it is not noted as "square acres" as students are used to seeing with respect to area measurement. Therefore, this question would require converting units before determining the dimensions of each piece.

> NOTE: If writing additional questions, keep in mind they...
> - Arise out of students' misconceptions.
> - Cause the student to think more deeply about the mathematics.
> - Should be answerable by the teacher.
> - Should be answerable by more than a yes or a no.
> - Can be direct.

5. To practice providing feedback, review Ginny's thinking about the task. Discuss the following questions (in order):

    ❏ What do you like about her work?
    ❏ What misconception(s) did she have?
    ❏ What questions could you ask her to move her thinking forward?

NOTE: The teacher provided several feedback questions. Ginny then attempted the task again. Review her second attempt and discuss the same questions from above.

Task Extension: Ginny's work can also be used for error analysis with students as a follow-up to the task.

NOTE: Tangrams can be cut from die cuts and foam, cardstock, construction paper, etc. Alternatively, students can make their own set of tangrams through paper folding and tearing/cutting. This is a good spatial reasoning exercise. Directions are below.

Paper Folding a Tangram

❏ Square up a piece of 8.5″ × 11″ paper. Fold the paper so a shorter side lies on tops of (coincides) with one of the longer sides. Fold back and forth, creasing each time, and tear off the rectangle. You should now have a square piece of paper and a rectangle. Keep the square.

❏ Fold the square along one diagonal and crease to make two congruent right triangles. Fold back and forth, creasing each time, and tear apart the right triangles. Set one aside.

❏ Take one right triangle and fold in half so you form another set of two congruent right triangles. Fold back and forth, creasing each time, and tear apart the right triangles. Set these aside. They are the first two tangram pieces.

❏ Take the second large right triangle and position it so the right triangle is at the top and the hypotenuse is the base. Fold the right angle (the square corner) down to the middle of the opposite side (the hypotenuse). Fold back and forth, creasing each time, and tear apart the small triangle on top from the isosceles trapezoid on the bottom. Set the triangle aside. This is the third tangram piece.

❏ Turn the isosceles trapezoid so the longer base is on the bottom. Fold the left side of the trapezoid over on top of the right side so you have folded it in half along a vertical line of symmetry. Unfold and fold the left bottom corner to the middle fold line so the bottom sides lie on top of each other. Crease well and tear apart the small triangle and the remaining small square on the left of the fold line. These are the fourth and fifth pieces of the tangram.

❏ Take the remaining trapezoid and turn it so the right angles are on the left and the longer base is on the bottom. Take the upper left corner (at the obtuse angle) and fold it down and to the left corner (the lower right angle) so the bottom sides lie on top of each other. Crease well and tear apart the small triangle on the left and the remaining parallelogram. These are the sixth and seventh pieces of the tangram.

Selling the Farm Tangram Solutions

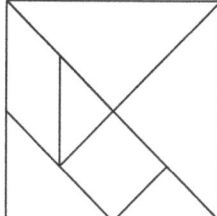

Tangram Square

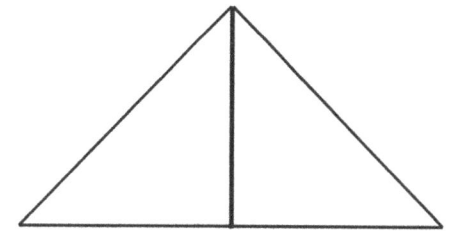

2 large triangles $= \dfrac{1}{2}$ the tract of land

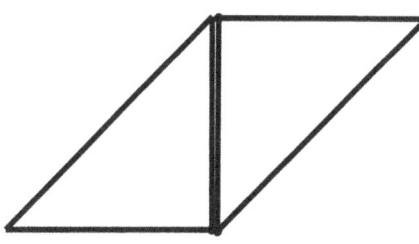

2 small triangles = 1 parallelogram

$2 \times \$525 = \$1050$

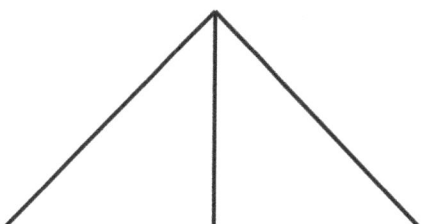

2 medium triangles = 1 large triangle

1 medium triangle $= \dfrac{\$2100}{2} = \$1050$

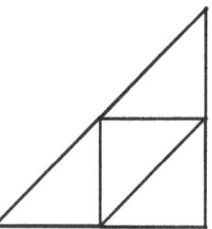

4 small triangles = 1 large triangle

2 squares = 1 large triangle

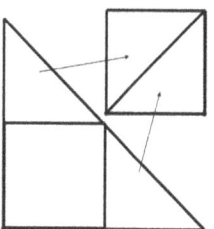

The two outside small triangles also make a square.

1 small triangle $= \dfrac{\$2100}{4} = \$525$

1 square $= \dfrac{\$2100}{2} = \$1050$

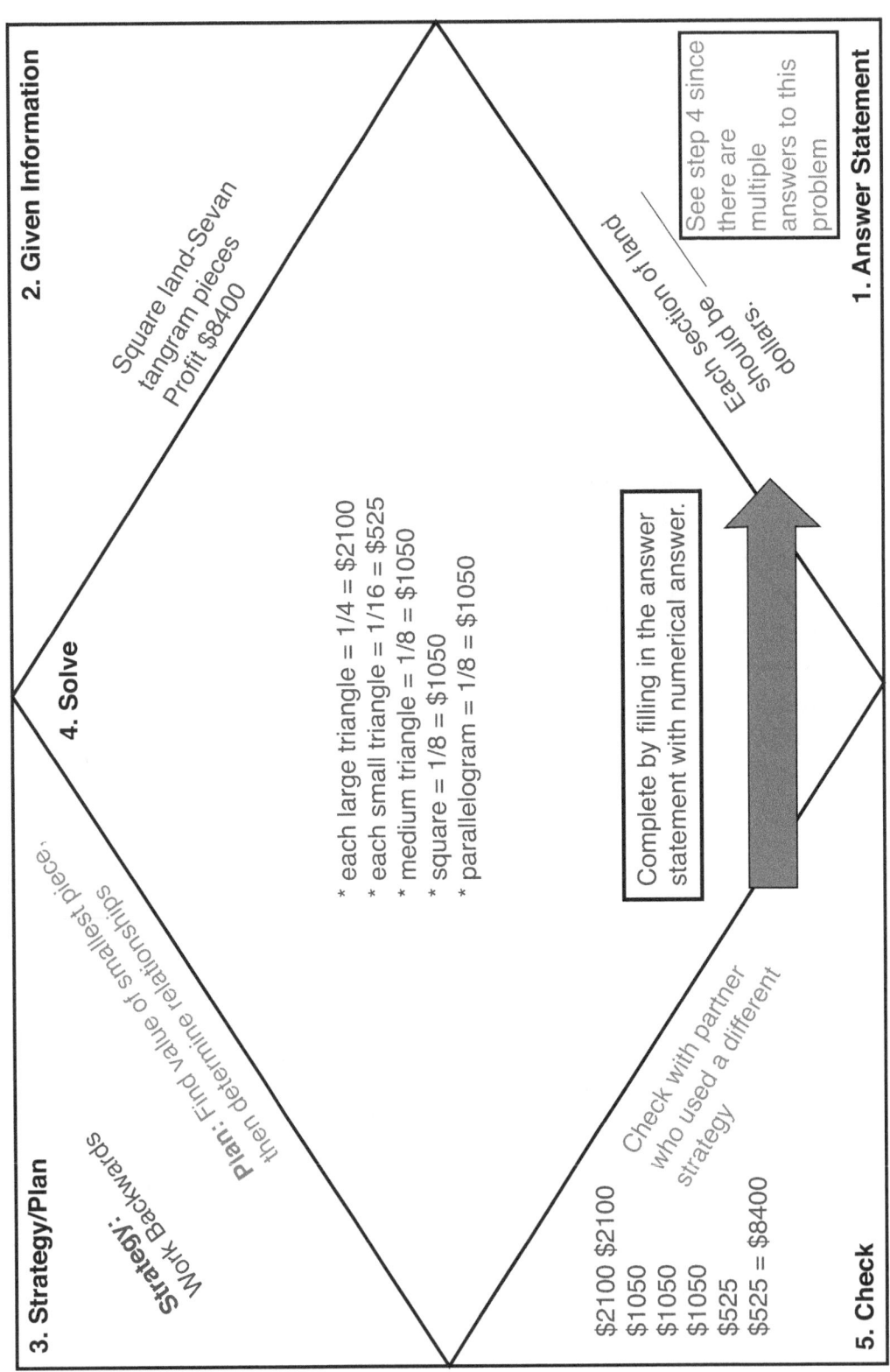

2. Given Information

Square land-Sevan
tangram pieces
Profit $8400

4. Solve

* each large triangle = 1/4 = $2100
* each small triangle = 1/16 = $525
* medium triangle = 1/8 = $1050
* square = 1/8 = $1050
* parallelogram = 1/8 = $1050

3. Strategy/Plan

Strategy:
Work Backwards

Plan: Find value of smallest piece,
then determine relationships.

1. Answer Statement

See step 4 since
there are
multiple
answers to this
problem

Each section of land
should be _____
dollars.

Complete by filling in the answer
statement with numerical answer.

5. Check

$2100 $2100
$1050
$1050
$1050
$525
$525 = $8400

Check with partner
who used a different
strategy

Tangram Master

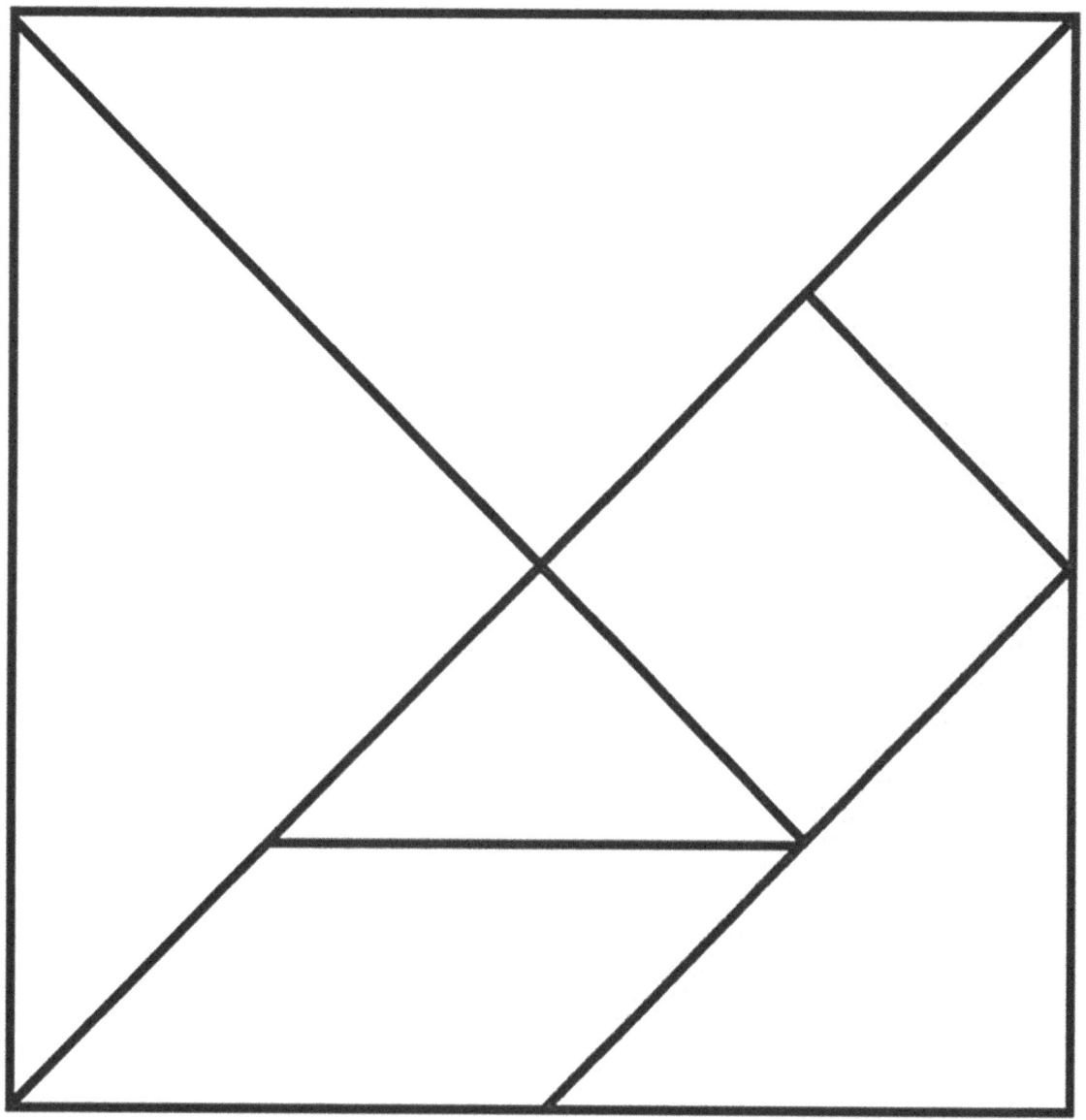

Selling the Farm WODB

A.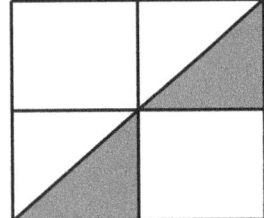

B.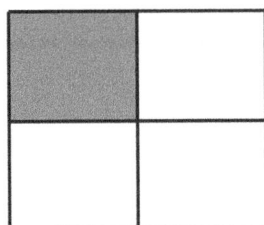

C.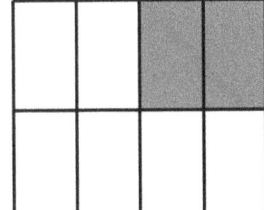

D.

Think-Write-Pair-Share	
Think about the problem. Farmer Jones is selling his land. It is the shape of a square. Mrs. Jones loves to play with tangrams. She convinced Farmer Jones to divide his land into seven pieces just like a tangram puzzle. Farmer Jones wants to make $8,400 when he sells all his land. How much should he price each section of the square? 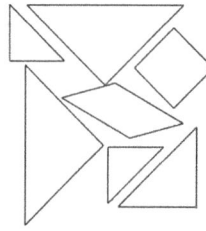 Tangram Pieces	**Write** by doing one of the following: If you can solve, choose a strategy and solve. If you cannot solve... ❑ Write the question that is being asked and/or create an answer statement. ❑ Write all facts you know about the problem. ❑ Write what additional information is needed or questions you have about the problem.
Workspace	
Pair with a partner and take turns discussing your strategies and solutions. Use this space to record strategies that were different from yours.	**Share** various strategies and solutions with the group. Use this space to record strategies that were different from those of you and your partner.

Think-Write-Pair-Share	
Think about the problem. Farmer Jones is selling his land. It is the shape of a square. Mrs. Jones loves to play with tangrams. She convinced Farmer Jones to divide his land into seven pieces just like a tangram puzzle. 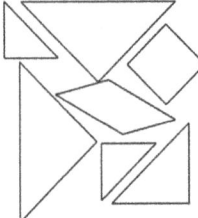 Farmer Jones wants to make $8,400 when he sells all his land. How much should he price each section of the square? Tangram Pieces	**Write** by doing one of the following: If you can solve, choose a strategy and solve. If you cannot solve... ❑ Write the question that is being asked and/or create an answer statement. ❑ Write all facts you know about the problem. ❑ Write what additional information is needed or questions you have about the problem.

Workspace

Five-Step Problem-Solving Process

3. Strategy/Plan   4. Solve   2. Given Information

5. Check   1. Answer Statement

Pair with a partner and take turns discussing your strategies and solutions. Use this space to record strategies that were different from yours.	**Share** various strategies and solutions with the group. Use this space to record strategies that were different from those of you and your partner.

Selling the Farm: We're Stuck

1. How does making the tangram square help in working this problem?
2. What is the relationship between each of the pieces of the tangram? How do you know?
3. What do you need to know to determine the value of each piece of land?
4. What is the relationship between each of the pieces of the tangram and the whole tract of land? How will this help determine the value of each?

Selling the Farm: We're Stuck

1. How does making the tangram square help in working this problem?
2. What is the relationship between each of the pieces of the tangram? How do you know?
3. What do you need to know to determine the value of each piece of land?
4. What is the relationship between each of the pieces of the tangram and the whole tract of land? How will this help determine the value of each?

Selling the Farm: We're Stuck

1. How does making the tangram square help in working this problem?
2. What is the relationship between each of the pieces of the tangram? How do you know?
3. What do you need to know to determine the value of each piece of land?
4. What is the relationship between each of the pieces of the tangram and the whole tract of land? How will this help determine the value of each?

Selling the Farm: We're Stuck

1. How does making the tangram square help in working this problem?
2. What is the relationship between each of the pieces of the tangram? How do you know?
3. What do you need to know to determine the value of each piece of land?
4. What is the relationship between each of the pieces of the tangram and the whole tract of land? How will this help determine the value of each?

Selling the Farm: We're Done

1. How can three different geometric shapes of land have the same value?
2. Why did the values of each piece of land calculate to be whole numbers?
3. How would the values change if the total price were not a multiple of the denominators of the fractional pieces?
4. If the area of the tract of land measures 100 acres, what would be the area of each piece of land? What would be the dimensions of each piece?

Selling the Farm: We're Done

1. How can three different geometric shapes of land have the same value?
2. Why did the values of each piece of land calculate to be whole numbers?
3. How would the values change if the total price were not a multiple of the denominators of the fractional pieces?
4. If the area of the tract of land measures 100 acres, what would be the area of each piece of land? What would be the dimensions of each piece?

Selling the Farm: We're Done

1. How can three different geometric shapes of land have the same value?
2. Why did the values of each piece of land calculate to be whole numbers?
3. How would the values change if the total price were not a multiple of the denominators of the fractional pieces?
4. If the area of the tract of land measures 100 acres, what would be the area of each piece of land? What would be the dimensions of each piece?

Selling the Farm: We're Done

1. How can three different geometric shapes of land have the same value?
2. Why did the values of each piece of land calculate to be whole numbers?
3. How would the values change if the total price were not a multiple of the denominators of the fractional pieces?
4. If the area of the tract of land measures 100 acres, what would be the area of each piece of land? What would be the dimensions of each piece?

Student Sample Attempt #1

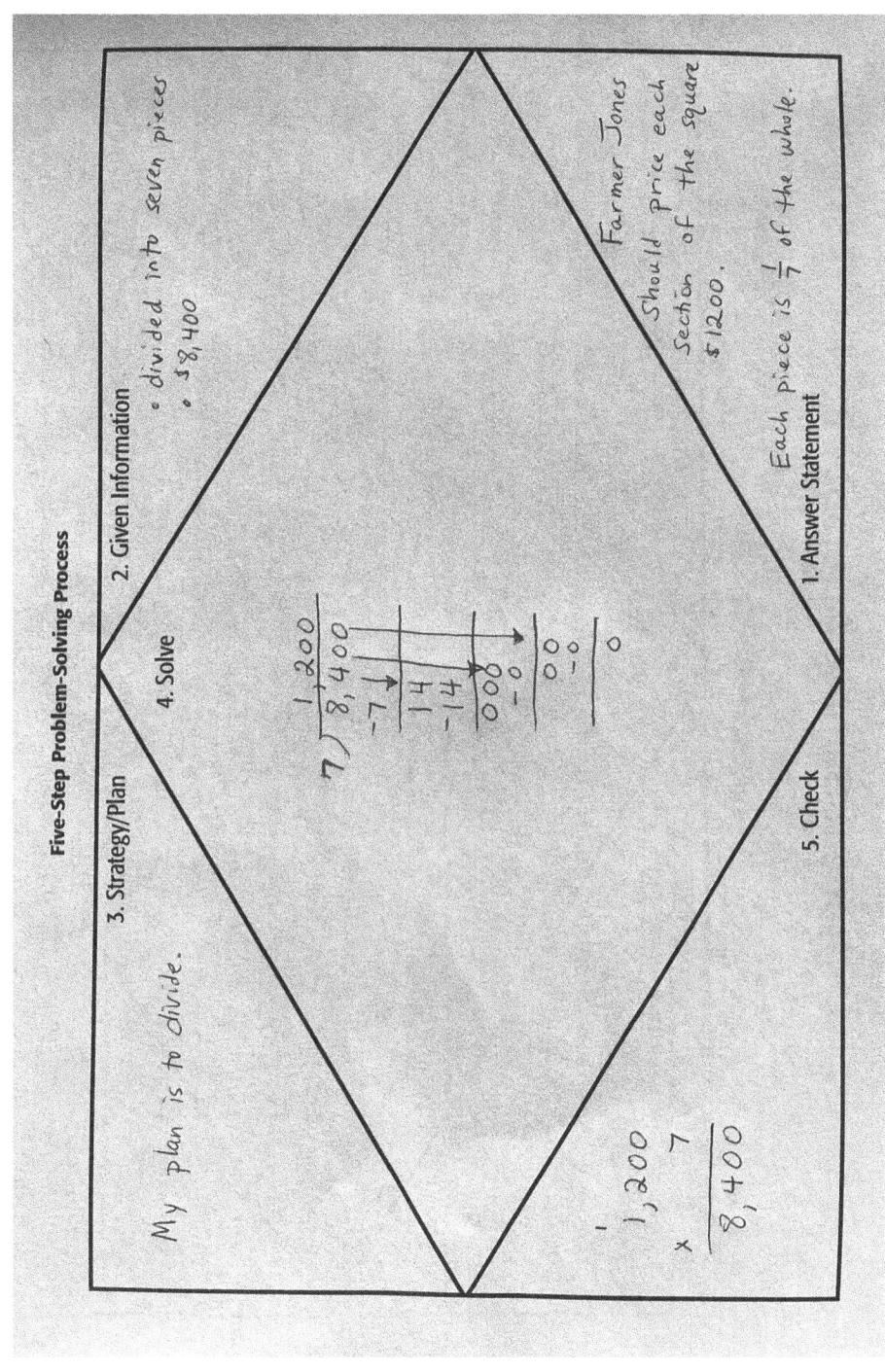

Five-Step Problem-Solving Process

2. Given Information

- divided into seven pieces
- $8,400

1. Answer Statement

Farmer Jones should price each section of the square $1200.

Each piece is $\frac{1}{7}$ of the whole.

3. Strategy/Plan

My plan is to divide.

4. Solve

$$7\overline{)8,400}$$

5. Check

$$\begin{array}{r} 1,200 \\ \times\ 7 \\ \hline 8,400 \end{array}$$

Student Sample Attempt #2

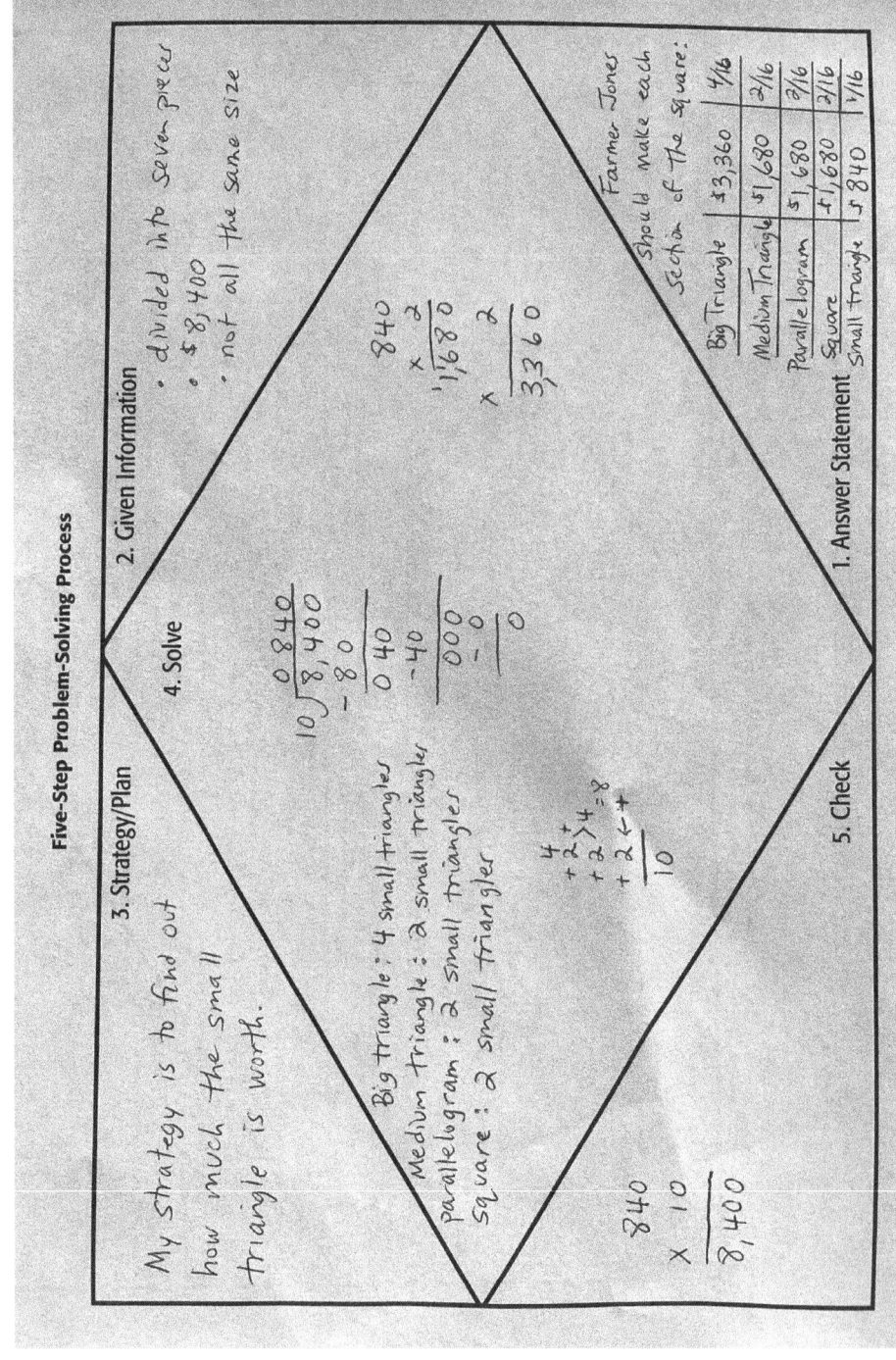

Five-Step Problem-Solving Process

2. Given Information

- divided into seven pieces
- $8,400
- not all the same size

3. Strategy/Plan

4. Solve

My strategy is to find out how much the small triangle is worth.

$$840$$
$$\times 2$$
$$\overline{1,680}$$

$$\times 2$$
$$\overline{3360}$$

$$\begin{array}{r} 0840 \\ 10\overline{)8,400} \\ -80 \\ \hline 040 \\ -40 \\ \hline 000 \\ -0 \\ \hline 0 \end{array}$$

Big triangle: 4 small triangles
Medium triangle: 2 small triangles
Parallelogram: 2 small triangles
Square: 2 small triangles

$$\begin{array}{r} 4 \\ +2 \\ +2 \\ +2 \\ \hline 10 \end{array} \quad 4=8$$

1. Answer Statement

Farmer Jones should make each section of the square:

Big Triangle	$3,360	4/16
Medium Triangle	$1,680	2/16
Parallelogram	$1,680	2/16
Square	$1,680	2/16
Small triangle	$840	1/16

5. Check

$$840$$
$$\times 10$$
$$\overline{8,400}$$

QUAD Reflection

Q: Question

What questions do you still have about the task?

U: Understanding

What do you now understand after working with this task?

A: Activate

How did this task activate you as a learning resource for a peer or a peer for you?

D: Discourse

What discourse did working this task prompt?

QUAD Reflection

Q: Question

What questions do you still have about the task?

U: Understanding

What do you now understand after working with this task?

A: Activate

How did this task activate you as a learning resource for a peer or a peer for you?

D: Discourse

What discourse did working this task prompt?

Afterword

We have always had educators ask us if they could get a list of the questions we used during the day's training, a list of the writing prompts we used, an example of a Differentiated Instruction (DI) strategy we mentioned, etc. Our questions came from our interactions with the participants; the writing prompts, while mostly intentional, also might have been inspired by a comment or question from a participant, and, yes, we did have examples of the various strategies that we often mention. At the same time, our editor, Lauren, to whom we owe so much thanks and gratitude, asked us if we would ever consider doing a book of consumables for educators to use. Hence, the seed was sown that finally grew into what became this book series.

Everyone acknowledges that communicating in mathematics is essential. Communication was one of The National Council of Teachers of Mathematics's original five process standards. Over the years, we have collected, found, and created many classroom resources that provide authentic opportunities for students to communicate mathematically. So, the next step was culling through the many resources we had used and developed over the years. We sorted, resorted, looked at, accepted some, rejected others, and even created new ones as needed. We wanted to offer this series in four books to meet the individual needs of the various grade bands. At the same time, we wanted to provide examples and prompts that would cover the breadth of the grade band's mathematical topics and provide materials to support deepening the understanding of the topics.

Then to the writing and pulling together of the resources, which for us as mathematicians, the latter was much less challenging. This is ironic since this book focuses on writing and communicating in mathematics! What emerged is what you have on these pages. We hope that this resource becomes a go-to to meet your everyday classroom needs for providing opportunities for your students to engage in communicating about mathematics and communicating mathematically. Take what we have provided, expand on it, and make it your own. As you reimagine, retool, and even create your versions, we ask that you reach out and share. We would love to hear from you. You can find us at https://tljconsultinggroup.com/about-us/tammy-jones/ and https://leslietexasconsulting.com/.

Bibliography

Growney, J. *Intersections—Poetry with mathematics.* https://poetrywithmathematics.blogspot.com

Jones, T. L. and Texas, L. A. (2017). *Strategic journeys for building logical reasoning: Activities across the content areas.* New York: Routledge, Taylor & Francis Group.

Smith, M. S. and Stein, M. K. (2022). *5 Practices for orchestrating productive mathematics discussions.* Reston, VA: National Council of Teachers of Mathematics, Inc.

Southern Regional Education Board. (2018). *Making math matter: High-quality assignments that help students solve problems and own their learning.* SREB. https://www.sreb.org/sites/main/files/file-attachments/18v04_math_matters_report_final.pdf?1521473373

Texas, L. A. and Jones, T. L. (2013). *Strategies for common core mathematics: Implementing the standards for mathematical practice.* New York: Routledge, Taylor & Francis Group.